CULTURE AND POLITICS
OF
SUSTAINABLE WATER MANAGEMENT

CULTURE AND POLITICS OF SUSTAINABLE WATER MANAGEMENT

Dr. R.N. Sahay

Oxford Book Company
Jaipur ▪ New Delhi

Published By

Oxford Book Company

Regd. Off. "Bony Residency", Gate No. 2
Opp. Tilak Public School, Vishweswariya Nagar,
Gopalpura Road, Jaipur - 302018, India
Ph: 0141 - 2761280, Telefax: 2761381
email - oxfordbookjpr@gmail.com
website - oxfordbookcompany.com

Delhi Office: 102, 1st Floor, Satyam House,
4327/3, Ansari Road, Darya Ganj,
New Delhi - 110002, India
Ph: 011-45652440

Distributed By
KEYA INTERNATIONAL PRESS LLP
E-mail: keyainternationalpressllp@gmail.com

Published in India, 2025

ISBN : 9789391096199

Typeset by Shivangi Computers, Jaipur

Printed and bound in India by Thomson Press Ltd.,
New Delhi

Contents

Preface

Scientists predict that erratic precipitation patterns will persist, producing recurring droughts in years to come; continued growth in demand will only exacerbate the consequences of weather shifts. For policymakers, water shortages present three central challenges. First and most obviously, scarcity reminds us of the limits of our natural resources, and forces us to consider conservation measures. Second, scarcity requires that we assess the efficiency of water allocation and usage. Third—especially in the context of water, a universally necessary resource—equity demands that basic needs be met, and that one group of users not be allowed to exclude another group from an essential resource. In recent decades, water shortages have compelled eastern lawmakers to face these three issues—conservation, efficiency, and equity—directly. The result has been growing dissatisfaction with the legal regime that had historically governed the distribution of water rights in the East: common law riparianism. Under riparianism, "an owner of land abutting a waterbody has the right to have the water continue to flow... subject to the equal rights of each owner to make strictly limited use of the water." Riparian landowners, also called riparians, "share the common benefits that arise from adjacency to defined bodies of water."

Although riparian law had governed surface water in the East from the time of the first Anglo-American settlers, it came under increasing criticism during the latter half of the twentieth century. From a conservation standpoint, riparianism failed to protect against excessive diversions by riparians, or to ensure minimum stream flows for the public. From an efficiency standpoint, riparian rights were inherently uncertain because they were correlatively defined, and thus could shift over time as neighbouring users and uses changed; uncertainty, in turn, inhibited investment and prevented the development of markets in transferable water rights. And from an equity standpoint, riparianism was accused of favoring private agrarian interests.

Moreover, because rights were determined by ex post litigation, riparianism—like any common law regime—lacked a procedural mechanism for reallocating resources to those in need during times of scarcity. In the hopes of addressing these doctrinal shortcomings and better dealing with water shortages, approximately half of the eastern states formerly governed by common law riparianism established water permit systems, also known as permitting systems. These regulated riparian statutes—often adopted in response to scarcity concerns —envision a rationalized, administrative approach to water rights allocation. In theory, regulated riparianism requires that all direct users obtain a limited-duration, renewable permit from a state administrative agency.

The present publication has been designed to approach this subject in a simple and lucid style. The method of presentation is very clear and lucid, which can be easily followed by the students.

—Dr. R.N. Sahay

1

Elements for the Modern Era in Basin-wide Problem Solving

The allocation, management, and development of water in the West are dynamic. Yet the basic law governing allocation of western water has not changed substantially in over one hundred years, and is steeped in archaic concepts fashioned to address situations no longer relevant. Why this dichotomy between reality and the law? In short, the resource itself, our demands on it, and our view of its value are changing at a rate that outpaces the law's ability to' adapt. Supply fluctuates on both a seasonal and a long-term basis. Estimates indicate that 1.2 billion people globally experience a shortage of potable water, and given current population trends that number will only increase.

Many Indian reservations in the United States, like the developing world, lack potable water. In the West, growing urban demand, recent recognition of tribal rights, and needs for critical habitat place increasing strain on this scarce resource. To meet these needs in the face of growing concern with the cost of water development on the integrity of ecosystems and the federal pocketbook, the focus in water policy both globally and in the West has moved away from the twentieth century emphasis on water development and toward improvements in management and efficiency. Yet many of these efforts in the western United States run headlong into the archaic law governing water allocation. Western water policy makers and practitioners find themselves in a constant struggle to introduce sufficient flexibility into the law to address changes in supply, demand, and values.

In the effort to address modern problems, negotiation plays an increasingly important role. Collaborative processes aimed at resolving local issues are taking place in many of the water basins in the West.

Frequently, what began as a focused process to settle, for example, tribal water right claims, has expanded to cover basin-wide issues. This expansion is a natural result of the fact that despite jurisdictional barriers, use of water in one part of a basin affects its availability in other parts. The current ad hoc approach to settlement has given rise to a variety of processes, thus providing a fertile ground for testing concepts for change in water dispute resolution.

The growing use of negotiation to solve problems not adequately addressed by existing law may very well herald a new era for water distribution and management in the West, one tailored to the problems faced by specific water basins and structured around governance that mimics basin boundaries. Two of the measures agreed to in the Milk River negotiations—establishment of an intergovernmental committee to coordinate the management of water across jurisdictional boundaries and development of a program to bank water for redistribution during drought—are major steps in efforts to introduce both basin-wide governance and flexibility in water management. In addition, the Milk River negotiations reversed the inequity brought on by federal emphasis on development of water surrounding an Indian reservation at the expense of tribal water rights.

West to the Great Basin, where the threat of reallocation of water to meet the needs of endangered species and growing urban needs in the Truckee River Basin of California and Nevada is giving rise to a negotiated plan governing operation of the heavily developed river. By introducing flexible management to existing infrastructure, the Truckee River negotiations are overcoming substantial barriers to reallocation of water.

Because efforts to alter river governance, reallocate water, and reverse environmental harm to riparian systems are taking place on most major water basins in the West, this Chapter evaluates the efficacy of the negotiation process in this context and seeks a framework within which the results of local negotiation processes can be evaluated, recommending a process to assure that the broader interest in the fair and efficient use of water resources and the durability of the solutions are not overlooked. The processes used and the issues addressed on the Milk and Truckee Rivers are sufficiently different to provide a broad spectrum for this analysis. The negotiated settlement of tribal water rights in the Milk River Basin occurred within the context of a statewide general stream adjudication and resulted in the establishment of a basin-wide water coordinating committee and joint federal-state-tribal mechanisms for water distribution and dispute resolution.

The Montana process brings only governmental entities to the table, but includes extensive public involvement and interdisciplinary problem solving to arrive at solutions. Finalization of agreements requires legislative, congressional, and Tribal Council (or tribal referendum) approval in addition to entry of the agreement in a Montana Water Court decree. In contrast, negotiation of interstate water allocation, new basin-wide storage operating criteria, and modifications to storage allocation in the Truckee River Basin of Nevada and California occurs within a process established by Congress designed to resolve years of litigation and earlier failed attempts at negotiation. Federal representatives participated at a higher level and played a stronger role in achieving resolution than in the Milk River negotiations.

Also, unlike the Milk River negotiations, public involvement is minimal until a final agreement is reached; however, governmental entities at the table strive to represent the public interest. Participants at the table include all major water users and managers of stored water, whether private or governmental. Finalization of agreements requires approval by governmental parties specified by Congress as necessary. Both the Milk and the Truckee River Basin processes resulted in agreements that will substantially improve water management, allocation, and use. Yet the results of both negotiations have flaws. The opportunity to learn from the achievements and failings of these processes and apply them to other basins provides the impetus for this series of Articles. These criteria include evaluation of the efficiency and fairness of the process itself, as well as the durability of the outcome. Efficiency relates to the cost of resolution; fairness concerns the allocation of the benefits of use of the water resource; durability pertains to the sustainability of both the institutions established to manage water and enforce water rights, and to the environmental health of the riparian system.

Nevertheless, the Milk River process may not result in a sustainable use of the water resource because solutions remain heavily reliant on state and federal subsidy and may do nothing to restore the environmental integrity of the aquatic system. In contrast, due to a shifting of power through litigation, negotiation on the Truckee River took place against a backdrop that gave equal voice to restoration of aquatic health. Nevertheless, institutions established to implement the Truckee River agreement may not provide sufficient flexibility to adapt to future change in water supply and demand, and the absence of a key party from the table during final negotiations may render the solutions vulnerable or at least more difficult to implement.

Finally, looks at potential problems with the current ad hoc, locally driven approach to negotiation of basin-wide water issues and provides recommendations for key elements necessary to assure accountability to a broad range of interests.

In addition, recommends congressional criteria for approval of water settlements and authorization of federal funding to implement them to promote fair allocation of the benefits of use of the water resource and movement toward sustainable use of the resource, and the use of federal subsidies only to these ends. These recommendations attempt to span the gap between local and national objectives, between the reality that water use is local and drives local economies and the fact that the continued viability of our water resources and the legacy we leave to future generations in water infrastructure, social stability, and environmental amenities is truly national in scope.

Growing frustration with the inability of current institutions and legal rules to solve modern problems of water allocation and management in the West is driving people to the negotiation table in an effort to fashion a new approach to resolution of water issues. This section examines the types of problems people seek to resolve in a basin-wide forum and why the process of negotiation is emerging as the dominant forum for their resolution.

The Problems

The problems facing the two water basins studied serve to illustrate the water issues giving rise to new approaches. Thus, it is useful to return to those two basins before discussing water issues in the West in general.

The Milk River Basin, Montana

Lack of a mechanism to coordinate water management among jurisdictions, inability to respond to drought, and underdevelopment of tribal water resources are three of the major issues that brought people together to seek solutions in the Milk River Basin of Montana. To summarize the jurisdictional morass: The Milk River has its headwaters in Glacier National Park in the United States, and becomes the Milk River on the Blackfeet Reservation where it receives water from the St. Mary River as part of the Milk River Reclamation Project, one of the first Reclamation projects approved in 1903 by the newly formed Reclamation Service.

The river then leaves the United States, crossing the Canadian provinces of Alberta and Saskatchewan before returning to the United States at what is referred to as the Eastern Crossing. "Downstream from this point, the river serves seven private irrigation districts and one United States Bureau of Indian Affairs (BIA) irrigation district, and flows past three Indian reservations, numerous Indian allotments, and two National Wildlife Refuges before it joins the Missouri River below Fort Peck Dam." Thus, four federal agencies, five Indian reservations, seven irrigation districts, one state, two Canadian provinces, and two countries manage or claim a right to water from the Milk River.

Three of the four reservations—Blackfeet, Fort Belknap, and Fort Peck—were created by an Act of Congress in 1888 out of a reservation encompassing the entire Milk River Basin in the United States. The Act opened the remaining area in the Milk River Valley to settlement now served by the Milk River Reclamation Project. Prior to development of the Reclamation Project, private farmers had started to divert water from the Milk River upstream from the Fort Bell(nap Reservation. Three years later, on July 5, 1898, agents of the federal government at Fort Belknap began diverting one thousand miner's inches (approximately 125 cubic feet per second (cfs)) of water from the Milk River for irrigation of reservation land. Insufficient water to serve both diversions during drought left the downstream tribal diversion dry. The resulting lawsuit led to the landmark decision in Winters v. United States, recognizing the reserved water rights of the Fort Belknap tribes and setting the stage for reserved water right recognition for other Indian tribes.

Almost a century after Winters, no comprehensive tool for coordinated water management in the Milk River Basin exists, nor have the Tribes' reserved water fights ever been quantified. Further, despite the reclamation project, water use in the basin remains defined by drought. The United States Bureau of Reclamation estimates that water for irrigation from the Project is short in five out of ten years and that shortage will become even more pronounced when Canada and the Gros Ventre and Assiniboine Tribes develop their shares. Nothing in the allocation scheme defined by the doctrine of prior appropriation, which governs the distribution of non-reserved water rights in Montana, requires that water, during shortage, go to the most productive or least drought tolerant use. Instead, the water user whose ancestors developed the water first takes all.

Moreover, 94 years after the United States Supreme Court ruling in Winters, the water rights of the Fort Belknap Reservation remain woefully underdeveloped. Although originally designed to serve over 10,000 acres, the Fort Belknap Indian Irrigation Project serves less than 5,000 acres. Absent storage, natural flows on the Milk River during irrigation season are simply inadequate to supply water to the entire project. Tribal lands were among those authorized for service from the Milk River Project; however, the Fort Belknap Agency refused to participate.

The Truckee River Basin, California and Nevada

The Truckee River takes its water supply from the snowpack of the Sierra Nevada Mountains in California, and terminates in Pyramid Lake in the desert of Nevada. Along the way it serves kayakers, fishermen, hydropower stations, municipal needs, and a major diversion to the Carson River Basin for a federal reclamation project. To even out the cycle of flood and drought typical of rivers fed primarily by snowmelt, the Truckee River is regulated by five major federal reservoirs and several private reservoirs.

The terminus of the Truckee River, Pyramid Lake, is located within the Pyramid Lake Paiute Indian Reservation. In 1844, John C. Fremont observed that the lake and the mouth of the river teemed with Pyramid Lake cutthroat trout (Oncorhyncus clarki henshawi)—which is a stock of Lahontan cutthroat trout—and a sucker known as the cui-ui (Chasmistes cujus). Diversions of the river to satisfy the reclamation project lowered lake levels, blocking passage of fish to spawning grounds. The Pyramid Lake cutthroat trout disappeared entirely from Pyramid Lake in the late 1930s or early 1940s, though a similar strain of Lahontan cutthroat trout was subsequently introduced. Years of litigation attempting to reallocate water to Pyramid Lake ultimately upheld the dominance of appropriative water rights for irrigation. Only after passage of the federal Endangered Species Act (ESA) did the flow of the river begin to change.

The Fish and Wildlife Service listed the Lahontan cutthroat trout as threatened in 1975, and the cui-ui as endangered in 1967. Meanwhile, the urban areas of Reno and Sparks in Nevada grew, placing an increasing demand on water from the Truckee River Basin for municipal and industrial (M&I) needs. Despite efforts to obtain water rights for M&I use with an early priority date through purchase and transfer of irrigation water rights, the M&I water supply is insufficient during drought.

The West

The Milk and Truckee River Basins are indicative of the problems facing most water basins in the West. The Western Water Policy Review Commission identified the following among the water challenges facing the West:

1) Urban growth. Nine of the ten fastest growing cities in the United States are located in the West.
2) Unhealthy aquatic systems. The Biological Resources Division of the United States Geological Survey identifies freshwater fish as the single most endangered vertebrate group in the United States. In the Great Basin, nearly two thirds of the native fish are either listed under the ESA or considered of concern by the United States Fish and Wildlife Service. Water development is considered second only to the introduction of nonnative fish as a cause of these problems.
3) Poor water quality. The Environmental Protection Agency reports that over a third of the nation's rivers are use-impaired due to poor water quality. The water quality impairment is caused both by chemical pollution and physical alteration of streams.
4) Unfilled tribal water claims. Although reserved water rights arise at the time of establishment of an Indian reservation, development of tribal water has lagged far behind that of their neighbours. The presence of endangered aquatic species handicaps current attempts to develop tribal water in heavily appropriated basins.
5) Increasing periods of drought and flood due to climate change. Changes predicted to accompany global warming include increases in extreme weather conditions, reductions in crop yields in certain regions, and increased water scarcity in arid regions.

In the face of these changes in water supply and demand, and the recognition of environmental harm, westerners are beginning to alter the fundamental basis for allocation and management of water. This process of change is made difficult by the fact that the historic legal structure gave control of rater resources to those seeking to put them to economic use, and then locked that use in place by giving priority in order of date of development rather than current value of the use.

The historic system is complicated by the overlay of modern federal environmental regulation, which tends to reflect current values, is often directly at odds with consumptive use of water, and provides no clear guidelines on which use of water—human or the environment—should prevail. Inevitably the choice of directions taken to meet these challenges will be determined by the water supply itself.

In a vast area between the Mississippi River and the Sierra Nevada Mountains, rainfall is generally less than the twenty inches per year needed to farm without irrigation. Under the influence of aridity, westerners have a special relationship with the West's major rivers. These rivers control not only where people can live in the West, but also control where people want to live. John Wesley Powell, on surveying these arid regions in the late 1800s, recognized that the major rivers of the West would control its development. He recommended that the federal government eliminate the straight-line, rectangular survey so dear to the engineer, and draw property boundaries along topographic divides.

Aridity is not the only defining feature of this region. The water supply fluctuates on both a seasonal and yearly basis. As a result, native plant and animal communities have adapted to rely on a climate that oscillates between flood and dessication. What water users and managers in the West are beginning to face is the truth of Powell's observations. In reality, no amount of political gerrymandering or creative engineering can alter the fact that a water supply sets its own boundaries. The next step is to discover a process that allows diverse political jurisdictions and diverse needs dependent on a water supply already heavily developed to adjust to that reality. Negotiation is emerging as the most successful process.

The Approach

The emergence of negotiation as the most successful process for resolving many basin-wide water issues warrants discussion because it reveals some of the necessary elements of a process that will be effective. Reasons for resort to a collaborative process include: 1) the inadequacy of litigation for resolution of resource allocation problems, 2) ability to use the factual complexity of water supply and demand to expand availability and protection of the water resource, and 3) the need for participation by a broader range of interests.

The Inadequacy of Litigation

In general, people are unlikely to come to the table to negotiate solutions simply due to a sense of duty to community. Instead, people negotiate because "they seek to do together what they cannot do alone." Both the complexity of the water supply and the complexity of people's need for water cry out for a process flexible enough to accommodate that variability. In particular, as detailed in the following sections, water interests frequently come together because litigation is inadequate to address water management and distribution across jurisdictional boundaries, or because existing law governing water distribution within a jurisdiction is inadequate to address changing needs and values with respect to water use, even though the water supply and infrastructure itself is physically adequate to address those changes.

Multiple Jurisdictions

Water sources shared by multiple jurisdictions as described for the Milk and Truckee River Basins are the rule rather than the exception in the West. Variability in water supply is generally taken into account in administration of a court decree of water rights rather than the decree itself, and each jurisdiction will have its own institutions for administration and processes for resolution of disputes. On a shared water source, allocation problems remain between jurisdictions.

A onetime resolution of issues between jurisdictions will not do. Allocation of a water supply that fluctuates both seasonally and from year to year, and that may change over time due to climate change, requires that the jurisdictions involved with water management and allocation in a basin have an ongoing relationship. Attempts to litigate administrative schemes in the face of multiple jurisdictions tend to resolve only the issues at hand, ignoring the need to develop joint institutions that provide the forum for a continuing dialogue.

For example, consider the variety of legal outcomes in water basins shared between a tribe and a state. The Ninth Circuit has held that a tribe or the federal government has jurisdiction over administration, including change in use of its water rights, when the water source is located wholly within a reservation. However, the Ninth Circuit has also recognized state jurisdiction over non-Indian water use on a reservation when the water source is shared with private land. Under these scenarios, administration of water from the same source turns on the ethnicity of the water user. In response to a challenge to tribal dedication of a portion of the water right of the Wind River Reservation to instream flow, the Wyoming Supreme Court concluded that the change could only occur pursuant to state law.

The members of the Wyoming Supreme Court differed narrowly in their bases for the determination. One justice opined that only the State can hold an instream flow right under Wyoming law. This would present an absolute bar to the Tribes' desired use of their water. This limited ruling on ownership of an instream flow right was not necessary for the court to severely restrict the ability of the Tribe to dedicate water to instream flow.

Wyoming state law permits change in use of a water right only if the applicant proves no injury to other water users. Proving that an instream use imposes no greater impact than the agricultural use may prove impossible, because water quantified for the Wind River Reservation for agricultural purposes has not been developed. The Wyoming approach provides a simple rule: administration by a single jurisdiction. However, it reduces the tribal voice in use of its own water rights, and in this case also reduces the uses the Tribe may make of that water.

In this all-or-nothing approach, one or more governmental entity must lose control over distribution of its own water use rights, and the people using water under the losing jurisdiction have less voice in water management within their own community. Governments view loss of control over their own natural resources as an infringement on sovereignty. Negotiation allows development of a joint management scheme for a shared water resource in which all governments on the source have voices.

Through negotiation, governments retain control over the contours of any shared administrative forum. Negotiation provides a forum for the establishment of joint management and dispute resolution mechanisms that recognize the need to adapt solutions as water supply changes and allows participation by all entities affected on an ongoing basis. Unlike litigation, in which a solution is imposed by a third party, negotiation gives people both a voice in the process and responsibility for the outcome. Local involvement in decision making about the health and management of a water basin is key to successful implementation of solutions.

Existing Law Versus Changing Needs and Values

Increasingly, urban, habitat, and recreational needs are the dominant demand within western water basins. Not only are these needs the last to surface, but they are also the least able to tolerate shortage. The law does not care. Most water in the western United States is allocated on the basis of temporal priority: For private water rights, the priority date is the date of appropriation, and for federal and Indian reserved rights, the priority date is the date of establishment of the reservation.

The impact of allocation on the basis of temporal priority becomes apparent during drought, a frequent occurrence in the West where water supply fluctuates to such an extent that no year can be considered "average." Water shortage is not shared by water users relying on the same source. Instead, water rights are satisfied in order of priority. This approach was designed to protect the value of economic investment in water development at a time when the West was merely a source of resources for exploitation. To understand the inadequacy of this method of allocation for anything but purely economic interests, consider a bucket of water and a room full of thirsty people. If allocation of the water is governed by a simple rule under which the first person to arrive gets the first glass of water, the process of distributing the water is simple. Water is doled out to people in order of their arrival until it is gone. Of course, if the bucket is small, many people would remain thirsty and some might get water when they did not want it. Yet application of the priority system in litigation leads to this result.

Furthermore, litigation generally does not allow for prospective action. Parties must wait for an actual dispute to arise, timing that may be too late for the particular use or for the health of the aquatic system. Negotiation allows parties to anticipate future disputes and to design the institutions necessary to resolve them in a timely manner. Litigation runs the risk of providing a solution under a specific factual scenario of water supply that concerns only a part of the whole or may in fact never occur again. Yet despite this inflexibility in the law, there is considerable slop in the physical infrastructure built to supply water.

Many water supply systems built in the twentieth century are sufficiently overbuilt to allow for accommodation of new needs and values through changes in management and efficiency. These changes require cooperation across jurisdictional boundaries and creative solutions based on law and science that would be difficult to accomplish in litigation. Solutions that vary timing of water use, share shortage, alter and integrate the management of storage across jurisdictions, protect instream interests that may lack a legal right, and authorize and appropriate funding to improve water infrastructure and efficiency of water use may be possible by agreement where they would be foreclosed by litigation. Not only does negotiation provide the forum for agreement to resolve problems outside the shadow of existing law, but it also allows greater flexibility and application of the complex facts involved.

Handling Complex Facts

Certainly, courts can handle complex facts. However, where disputes are highly driven by facts such as a complex interplay between natural variables and human-created water development infrastructure, both diverse and variable water needs, and the need to design water allocation mechanisms in the face of a variable water supply, negotiation may provide better outcomes. First, in negotiation, scientific disciplines such as hydrology and municipal water supply engineering may take an equal seat at the table.

Where the adversarial system may see an irreconcilable conflict between a junior and senior water user during a drought, an agricultural engineer may see a difference in crop need that could accommodate both water uses.

Although scientists are involved as experts in litigation of cases involving complex facts, the role is quite different than the role in negotiation. Data developed for use in litigation often remain confidential unless disclosed at trial. As a result, opportunities for creative solutions remain hidden in those data. In contrast, participation by scientists and engineers in negotiation is often on an equal footing with legal and political representatives. This participation creates a greater likelihood that physical solutions will be discovered.

Second, active participation in problem solving by those with the expertise to evaluate the physical effects of creative solutions is necessary not only to expand the benefits from, or protection of, the water resource, but also to recognize and use the factual complexity of the problem to design flexible solutions.

This is not to suggest that science is a panacea. Legal over-allocation of water will not always be solved by changes in timing of use, new storage management, or efficiency improvements. But unless the complexity and limitations of the system are fully understood, effective means to allocate water in the face of that complexity cannot be designed.

Third, even when a complex water system cannot be fully understood with current data, negotiation allows jurisdictions to agree to an ongoing forum in which adaptive management and a dispute resolution mechanism can be employed to adjust to new variables. Of equal importance, the nonadversarial nature of the negotiation process may set the stage for improved relations between parties—a key component to successful implementation of any solution that requires a continuing relationship. This brings us to the people.

Broad Participation

Negotiation provides a forum for participation by a broader range of interest groups than are generally included in litigation. Historically, regardless of the forum, water development and allocation decisions were made by a narrow group of interests representing the legal rights to use water. Today, a much broader group, including those who use water in its natural watercourse without consumption, seeks a voice in water management and allocation decisions. Negotiation allows at least the possibility of participation by a broader range of interests. Although those holding the legal rights to water may view this participation as an annoyance, any resolution that leaves out major interests, whether holding legal rights or not, is incomplete. In the Truckee River Basin, foreclosure of the Pyramid Lake Paiute Tribe's pursuit of legal rights to water for fisheries led the tribe to pursue other avenues for water. By asserting the fiduciary duty of the United States to manage surplus water from a federal project in the interest of the Tribe, and by proceeding under the ESA, the Tribe made considerable inroads on traditional expectations for water allocation. For tribes in particular, negotiation provides a format for raising diverse water interests.

No substitute for water exists. Given a thirsty city, an endangered fish, or simply a high level of interest and a willingness to work the political process, those interested in water allocation, management, and preservation will not disappear simply due to absence of their name on a court decree of water rights. The sooner such interests are included in a process to resolve disputes, the more quickly water basins can move forward with comprehensive solutions.

Despite these arguments in its favour, negotiation has its critics. Whether the concerns raised warrant rejection of the process or can be accommodated in its design can only be answered by exploring the basis for its criticism.

Arguments Against Negotiation

Concerns over the use of a settlement process to resolve disputes fall into three categories relevant to negotiation to resolve water issues: 1) loss of the public value provided by precedent, 2) underrepresentation of the underprivileged, and 3) inability of the participants to make hard choices.

No Precedent

Critics of settlement argue that the public value of establishing precedent through litigation is substantial. In addition, concern is expressed that alternative dispute resolution will replace legal precedent with alternative solutions that lack a legal basis. These critics fear that application of alternative dispute resolution to environmental disputes "present[s] the danger that environmental standards will be set by private groups without the democratic checks of governmental institutions."

The concern over loss of precedent stems from the fact that although settlement serves the public purpose of resolving a particular dispute, no future parties in a similar situation can rely on or be guided by the outcome. While this statement is true, it is relevant only in the narrow legal definition of the term "precedent." First, as noted above, the ability to satisfy all water interests within a particular water basin often turns on the complex interplay of water supply and demand in that particular basin. Ability to do more than use solutions from one basin as ideas for possible application to another basin may not be necessary.

Second, governments and water users who establish a process for the fair and efficient resolution of disputes over a shared water resource are indeed establishing precedent for resolution of future disputes. As noted above, negotiation of water disputes in basins with multiple jurisdictions has had greater success in establishing mechanisms for joint water administration and joint dispute resolution than has litigation. Third, the value to future generations of improved relations gained through negotiation of disputes involving scarce resources is incalculable. In the wake of the tragic events of September 11, 2001, the keynote speaker at the annual Indian Water Rights Settlement Conference stated: I believe we are at a crossroads in our efforts to manage water in the West... [Them are] immensely difficult conflicts between endangered species, the traditional use of water and the often as yet to be quantified water needs of the Tribes. And of course, the continued dynamic between the need for new and innovative ways of managing water for a burgeoning population and decades of established water law and practice....

All the more reason, then, to use this extraordinary time to push for more civil discourse on these contentions natural resource issues, to advocate reconciliation with our native American neighbours, to do another one of the four things that Richard Harwood suggested citizens can do [in the wake of September 11th]: "Search out differing opinions. When under stress, we often seek solace from those who affirm our existing views. But now we must have the courage to actively seek out views that differ from our own. For it is only when we engage with different view points that we are forced to articulate our own views, examine deeply what we believe and reach a fuller understanding of what we value."

In short, precedent for adversarial relations between, for example, tribal and state governments, is precedent best lost. Concern that private interests will set standards for resource allocation if judicial resolution of disputes is avoided is best addressed by design of the process itself. Strong involvement by governmental entities in both the Milk and Truckee River Basin settlements provides a template for discussion of such mechanisms in subsequent sections.

Underrepresentation of the Underprivileged

Courts appropriately view one of their purposes to be protection of the rights of those who lack a strong political or economic voice. Thus, the application of alternative dispute resolution to public disputes such as those concerning water allocation raises concerns that those voices will find no willing ear. Consider first the traditionally underrepresented voice in resource disputes: that of the Native American tribes. While the tribes have won great victories in legal battles over water, there is an emerging view that within the framework of those victories tribes has greater control and greater ability to exercise their own sovereignty through negotiation rather than representation by the United States Department of Justice in litigation. Negotiation not only gives tribes control, it also allows them to raise indigenous values that may not be recognized as legal rights in court.

The challenge of representing the interests of a diffuse public in major water disputes is more problematic. Though the interest of the general public in management of water may not take the form of a water right, the interest of a community in the use and health of its local watershed, or the national interest in sustainability of scarce water resources, is hardly irrelevant.

Yet, the "public interest" lacks the legal framework of federal Indian law and the legitimacy and cohesiveness of a sovereign political entity to give it a separate and clear voice in negotiations. Processes currently underway for resolution of water disputes in the West represent a broad range of experiments on the issue of accountability to the general public. Satisfactory handling of this issue is key to design of a fair, efficient, and durable process.

Hard Choices

As the process of solving the water allocation and management problems in the West proceeds, hard choices remain. People do not rush to give up hard-earned economic rights such as those to use water. However, people have voluntarily made hard choices concerning water allocation in the past.

The doctrine of prior appropriation was not imposed by courts but adopted out of deference to custom in the mining camps. Though possibly enforced through the barrel of a gun, such custom represents a choice that it is best to have a clear rule to maintain peace over water allocation even if some go without water.

At the same time, the difficulty in malting these choices and the concern expressed above that those with the most economic or political clout will make them, cautions for retention of a strong judicial and legislative role in any process for resolution of water disputes.

Just as science cannot solve all water supply problems, negotiation cannot provide a forum for resolution of all disputes. But, given the potential gain from coming together and engaging in public discourse concerning the use and protection of western water basins, the following sections advocate negotiation as the dominant process while retaining roles for both the judiciary and Congress to ensure that negotiation moves in the appropriate direction and to provide a safety net should it fail.

In some cases, such as the Truckee River Basin, a preliminary round of litigation may be necessary to sufficiently alter the balance of power and bring those previously in control of the water resource to the table.

The Process of Negotiation: A Framework for Evaluation

Defining the Elements

The term "negotiation" can be applied to a broad spectrum of approaches to dispute resolution. Loosely defined, the term merely means that resolution of a dispute is reached through discussion among those interested without passing the decision-making authority to a neutral third party. Agreement to a particular solution is voluntary on the part of each party. Nothing in the term mandates a fair approach or good outcome. Given the broad range of alternatives encompassed by the term, it is important to establish a framework for both evaluation and design of negotiation processes adequate to resolve major water disputes. The following sections suggest a framework for the evaluation of the adequacy of a water negotiation process.

An adequate dispute resolution process consists of three elements: an 1) efficient and 2) fair process that 3) results in a durable outcome. The first two elements focus on process and the third on substantive components of the final resolution. Thus, deciding if a particular settlement process satisfies these elements requires analysis of both the procedure and the outcome. Each element is discussed in turn.

Efficient Process

Efficient processes achieve dispute resolution without excessive delay or cost. The vast amount of information and the large number of interests involved in any basin-wide water dispute render cheap, quick solutions unlikely. Efficiency of process therefore must be evaluated in comparison to other potential methods for achieving resolution. In the case of major water disputes, the principle comparison is between litigation and negotiation.

A properly designed negotiation process can result in a reduction in cost and time over litigation. The two most important features that aid efficiency in negotiation are: 1) reduction in cost of data collection through a combination of data sharing and joint data collection and evaluation, and 2) reduction in time through use of a less adversarial process. A decision to use negotiation does not guarantee inclusion of either feature. Rather, a conscious effort must be made to consider them in design of a process.

Less Duplication of Data—less Money

Understanding a complex hydrologic system with variable water supply, diverse water uses, and an overlay of human-induced changes to a natural system is not easy. It requires collection and review of large amounts of data. In a basin-wide water dispute where changes to existing infrastructure, management, or water use are contemplated, it may also be necessary to develop complex models to evaluate the effects of the changes. Litigation forces parties to develop such information on their own. Further, the adversarial process discourages data sharing and cooperative evaluation. Not only does this approach result in duplication of effort, but it also has two collateral effects that further delay and raise the cost of resolution.

First, inability to share information can cause parties to miss important options for resolution. Given the site-specific nature of water supply and use and the realization that considerable improvement in usable water supply can be made by fine-scale adjustments to existing water storage and distribution systems, it is clear that the best solutions will be found within the minutiae of technical and experiential information related to the specific water basin. Failure to collaborate on the collection and analysis of such information increases the likelihood that parties to the dispute will miss opportunities, particularly small-scale, low-cost opportunities.

Second, inability to collaborate on the collection and analysis of data guarantees that the initial argument in any process will be over whose data, analysis, or model is correct. This is true whether the process chosen is litigation or some method of alternative dispute resolution. In contrast, the joint data collection and analysis allows parties to negotiate from a common understanding of the technical nature of the problem, resulting in a savings of considerable time and, therefore, money. Thus, an efficient process rests on a foundation of joint data collection and analysis.

Less Adversity—less Time

In addition to saving time through potential use of joint data collection and analysis, negotiation saves time by eliminating some of the adversarial posturing inherent in litigation. The competitive nature of litigation can cause parties to avoid early disclosure of, or even lose sight of, their true underlying interests in the outcome. The result—similar to the failure to develop data jointly—is that parties miss opportunities for satisfactory solutions. Similar to the cost saving measures, reduction in adversity is not inherent in a negotiation process.

Negotiation simply provides greater flexibility to build means of reducing adversity into design of the process. Early disclosure of interests and avoidance of posturing and finger-pointing requires leadership among the parties, ground rules on negotiation, and, in some cases, assistance by a neutral third party. Subsequent sections will discuss building these features into the process, and address recommendations. Increasing process efficiency by reducing cost, time, and adversity, and thus providing a forum for a broader and less privileged group of participants, assists in rendering the process fair. However, additional features are necessary to assure that all interests have a voice in the process and a stake in the outcome.

Fair Process

A fair process in water dispute resolution is one that provides a forum for the airing and consideration of all interests including those of the public-at-large. Concerns raised with alternative dispute resolution, and in particular with settlement of environmental or natural resource disputes, generally relate to the inclusiveness of the process. Understanding the complexity associated with involving all interests requires recognition that there is both a community and a national interest in the use and health of water and water basins.

Prior to discussing the aspects of a process that would recognize these diverse interests, it is useful to divide them into three concentric categories: 1) water interests, 2) the community interested in the basin, and 3) the public-at-large. Each category is increasingly removed from the water resource. "Water interests" refers to those who use, manage, or benefit from water in the basin. This includes those who run a whitewater kayaking business on the water or rely on fisheries that depend on the water.

The "community interested in the basin" includes the water interests, but also incorporates those interested in the economy of the basin without direct ties to water, those who recreate in the basin, those in downstream basins who might be affected by changes in water use in the basin, and any other person who is interested, in the economic, social, or environmental aspects of the particular water basin. Finally, the "public-at-large" includes the community interested in the basin and the water interests, but also includes the general state, tribal, and national public.

Although a single individual may be a member of all three groups, the interests in each category are sufficiently different to warrant separate consideration. In addition, the interests in each group are increasingly difficult to represent in a dispute resolution process, and require separate consideration to avoid a solution that satisfies only local interests. The following sections conclude that addressing the concerns of each group requires that: 1) all water interests in the particular basin are present, 2) the concerns of the community interested in the basin, whether local or remote, are heard, and 3) the process is designed to be accountable to the public-at-large.

Water Interests

Commentators seem to agree that a fair water dispute resolution process must include all interested and affected parties. However, little is written about the fact that inclusiveness of all interests is important not only to the party potentially excluded but also to those remaining at the table. No comprehensive solution is possible, nor is any solution likely to endure, if an affected party has been excluded from the process. The implementation of solutions at the local level requires buy-in by those affected. Buy-in is unlikely to occur if people are excluded from the process. Because of the importance of water, an excluded party whose interests are not met will continue to pursue other avenues until their voice is heard.

A disadvantage of negotiation is that participation is voluntary. No mechanism such as subpoena power exists to force a party to the table. However, voluntary abstention from negotiation affects the durability of the outcome rather than the fairness of the process. An additional concern is that without court intervention, the underprivileged will not be afforded effective participation. Participation by the underprivileged requires not only a willingness by more privileged parties to include all water interests at the table, but a mechanism to fund effective participation.

As noted above, Native American tribes are historically underprivileged parties in water disputes. Effective participation by tribes requires funding for legal representation to provide a voice, technical representation to evaluate solutions, and joint efforts to generate and analyse data. The effectiveness of efforts by Congress and the Bureau of Indian Affairs to provide funding for tribes will be discussed under sections addressing recommendations. Regarding fairness, it is merely important to note that without a means to hind participation, an invitation for a tribe to come to the table is meaningless.

Basin Community

The avenue for participation by the community interested in the basin is public comment. This tends to occur in two different stages of a negotiation process. First, the public voice may be included through public comment on solutions as negotiators discuss them. Second, the public voice may be heard during the judicial or legislative process required to approve a final agreement. Each stage for public comment may be necessary to ensure a fair process. Postponing public comment until final approval can lead to the discovery of major problems after considerable resources have been spent on resolution.

In addition, it can result in missed opportunities for those who know the water supply to identify creative solutions. At the same time, judicial or legislative approval of an agreement tends to be more formal than general public comment and raises the likelihood that all comments will be heard and recorded. Moreover, comment in a judicial or legislative hearing is taken and acted on by someone with a more neutral view toward the negotiation. Comment at this stage may ensure that even the most divisive voice is heard.

Public Accountability

How to include the more diffuse interests of the public-at-large may be the most difficult problem raised by the trend toward local resolution of basin-wide water disputes. Clearly the entire population of living generations, not to mention future generations, cannot take a seat at the table or provide public comment. However, the importance of water beyond the local community currently living within a particular water basin warrants an effort at broader representation. One scholar argues for a moral check on accountability asserting that those at the table should consider effects on third parties or the larger society voluntarily. Water disputes are too complex for this approach. Parties at the table have enough difficulty ascertaining their own interests and the affects on them without trying to guess those of a third party. It is most likely that parties will look to governmental entities at the table to represent the public interest; however, adequate representation requires more specific guidelines than a moral check.

At the same time, a balance must be struck between the need to represent the larger public interest and the need to maintain flexibility in the development of a basin-specific solution. Two methods are apparent for giving voice to the larger public interest. Each will be discussed in light of the need to strike that balance. First, the process may accommodate the public interest by ensuring that the parties at the table negotiate an outcome that complies with standards set at the national level. Thus, adherence to water quality standards promulgated under the Clean Water Act and requirements of the ESA ensure that a national agenda aimed at maintaining a healthy water supply and ecosystem plays a role in the outcome. Interpretation of the application of national standards to the specific basin can be accomplished through participation of appropriate representatives of the federal government.

This approach is useful for issues that lend themselves to national standards. However, resort to negotiation is generally driven by the perception that a basin specific solution is necessary. National standards for water allocation issues other than water quality or endangered species are unlikely to address the details of a local solution and if modified to do so would thwart the very site-specific solutions that are proving successful. Second, interests of the public-at-large that do not lend themselves to national standards may be represented by governmental entities. Negotiations studied of this series on the Milk and Truckee River Basins included representation of statewide interests by the state governments and tribal interests by tribal representatives.

These governmental entities frequently used a process for public comment to determine the public interest. However, state, local, and tribal representatives cannot appropriately represent the national interest. Except for its adherence to federal standards, the current federal role in negotiations involving tribal water rights such as those on the Milk River tends to be narrow, focusing on its role as trustee in representing tribes, or its proprietary interest in a federal project. Designing a process to allow the federal representatives to identify national interests within a basin, while maintaining the integrity of the basin-specific interests the federal government must represent (such as those of the tribes and of federal reclamation projects), will be discussed in subsequent sections recommending means of altering current processes. The approach recommended for federal participation is equally applicable to state, local, and tribal governments struggling to identify the interests of their public.

Each of these approaches is useful in the design of a fair process. Most important is the initial recognition by the parties at the table that the voice of each of the circles of interest, including the public-at-large, is legitimate, that public input is not merely a procedural step, and that solutions will be both better and more durable if the public interest is taken into account. This leads to the third element of an adequate process: durability of the outcome.

Durable Outcome

Key to the viability of a process for resolution of water disputes is the likelihood of achieving a durable outcome. Processes designed to save money and time that include everyone accomplish little if the focus on process comes at the expense of a quality substantive outcome. "Durability" should not be confused with "finality." A key difference between litigation and negotiation is that negotiated solutions can recognize that the variability and complexity of water supply and demand sometimes render a "final" resolution of all water issues impossible and, more importantly, undesirable. Settlement must set the stage for a continuing dialogue to address changes efficiently and fairly when they occur. Instead of finality, durable outcomes should achieve: 1) a comprehensive solution, 2) equitable distribution of use of the water resource both within the current generation and between current and future generations, and 3) relative ease of implementation, including institutions for a continuing dialogue among those sharing the water resource.

A Comprehensive Solution

A durable outcome is comprehensive because it includes all interested parties and all major issues. First, just as inclusion of all interested parties is essential to the fairness of a negotiation process, it is also essential to the durability of the outcome. It is in the interest of each party to include all other interested parties to avoid subsequent erosion of a solution or barriers to implementation of a solution because parties excluded from negotiation attack through other avenues. In the Truckee River Basin negotiations covered in Part II of this series, a key party—the Truckee-Carson Irrigation District (TCID)—refused to participate in the settlement. Parties who remained at the table believe the first attack on the final agreement will come from TCID.

Second, comprehensive coverage of the issues, at least those that are interrelated, is essential to a durable outcome. The Western Water Policy Review Advisory Commission considers a comprehensive solution to cover water quality as well as quantity, surface and groundwater, human needs, and the environment. The complexity of a settlement this broad may render it impossible to achieve in a single agreement. A "comprehensive" solution need not solve all water related problems in a single step. However, at any one stage it is important to include all interrelated issues to avoid a solution that cannot be implemented.

For example, if federal water quality standards can be met under certain scenarios of water allocation and not under others, water quality must be considered in choosing options for water allocation. By establishing basin-wide forums, for continuing resolution of disputes and to implement solutions during initial efforts to solve basin-wide problems, negotiators can set the stage for additional steps to reach a comprehensive solution over time.

Equitable Distribution of use of the Water Resource

Equitable distribution of use of the water resource requires solutions that satisfy both intragenerational and intergenerational equity. First, equity within this generation requires that decisions on water distribution fairly distribute the rights to enjoy the use of the water, and allocate water according to current values. Second, equity between generations requires a healthy aquatic and riparian ecosystem, a viable water infrastructure, and institutions flexible enough to adapt to changing circumstances to be passed to future generations. These requirements for intergenerational equity are reflected in the concept of "sustainable use." Solutions that may aid in achieving the requirements of intragenerational equity—a fair distribution of the rights to enjoy use of water and allocation according to current values—will be discussed first, followed by a discussion of sustainable use and its relation to intragenerational and intergenerational equity.

Intragenerational Equity

The suggestion that a durable solution requires a fair distribution of the right to enjoy the use of water and allocation of that water according to current values is not intended as a radical social restructuring of water allocation schemes. To be "fair," solutions must acknowledge existing water rights under both federal and state law. At the same time, solutions must account for changing needs. In terms of western water, this balance can be struck if an agreement includes the following elements: 1) means to reallocate water based on changing need and value that fully compensate any lost right to water (taking into account that there is generally not a right to waste water), 2) provisions to address needs during drought, 3) measures to ensure water is not wasted for one use while another goes thirsty, and 4) provisions to ensure that these measures make water available to serve both economic and other uses.

Two avenues for reallocation of water to meet current needs and values are being developed throughout the West. First, marketing of water to provide either a drought water supply, or to move water from agricultural to urban needs, provides an avenue for reallocation in which the original water users receive sufficient benefit to either improve the efficiency of their water use or alter their economic base.

The context of the two case studies, a true market would ensure that water flows to its highest economic use, but may do nothing to repair environmental harm or distribute water to the underprivileged. Thus, mechanisms for retention of strong governmental control and measures to allow allocation to noneconomic uses are recommended. Second, reallocation has been imposed through enforcement of federal laws such as the ESA.

This second method is highly controversial and questions of whether the reallocation amounts to a Fifth Amendment taking, and the degree of certainty necessary regarding the water needs of the endangered species, have yet to be resolved. Nevertheless, the very lack of clear legal guidelines provides the uncertainty necessary to bring parties to the table. Consideration of a fair distribution of water includes serving the needs of the underprivileged. Indian water development has not kept pace with that of their neighbours. To provide a fair distribution of water use in any basin where tribes hold a share in the water resource, the solution must remedy this inequity. Currently, the primary barrier to development of tribal water resources is not the inability to reach settlement, but the inability to fund development once settlement is reached.

Intergenerational Equity: Sustainable Use

Passing healthy ecosystems, viable water infrastructure, and adaptable institutions to future generations—all essential elements in obtaining an equitable distribution of use of the water resource are reflected in the concept of sustainable use. Sustainable use of water resources is much talked about in theory and poorly understood in practice. Sustainable use of water has been defined as 1) "'development that meets the needs of the present without compromising the ability of future generations to meet their own needs'" or 2) "use of water that supports the ability of human society to endure and flourish into the indefinite future without undermining the integrity of the hydrological cycle or the ecological systems that depend on it." Both definitions paint with such a broad brush that they fail to define the measure of sustainable use within a given water basin.

The western basins addressed in this series are already heavily developed. Thus, the question is not how to avoid altering a natural system, but how to alter a developed system to render it, and the ecosystem supporting it, sustainable. With this question in mind, the following sections attempt to parse these broad definitions into a form that can be used to judge the sustainability of negotiated solutions in a specific developed water basin. If the concept of sustainability is to reflect both environmental and human needs, it must incorporate a framework that assigns value to more than a purely natural system.

Goodland and Daly provide a useful framework to evaluate sustainability of a natural resource when the human element is an integral part of the ecosystem. To account for the need for development to sustain human life, they divide the legacy left to future generations into the following: "natural," "manufactured," and "human" or "social" capital. "Natural capital" is defined as "our natural environment." "Manufactured capital" is the human created infrastructure.

Finally, "human or social capital" is the people and the institutions, information, knowledge, and culture they create. The categories are not interchangeable. Instead, a sustainable level of each type of capital must be maintained. Thus, use of nonrenewable resources requires equal investment in maintaining or developing their substitutes. Use of a renewable resource, like water, requires that the use not degenerate the value of the resource.

For example, the huge water projects built in the western United States in the early twentieth century by previous generations have given this generation cheap hydropower and less dependence on fossil fuels. Nevertheless, the development occurred at considerable cost to our natural legacy.

By completely altering the flow of rivers, natural capital was sacrificed for manufactured capital. Some conversion of natural to manufactured capital is necessary to achieve the quality of life we enjoy in the United States. A sustainable solution need not eliminate water infrastructure. In fact, it must not do so if it is to meet the needs of current generations. However, where the conversion of natural to manufactured capital becomes unsustainable is in the failure to account for external environmental costs in water development. To remedy this failure, modern durable solutions must both repair existing damage to natural capital and minimize future damage.

Sound water infrastructure that minimizes damage to natural capital and efficiently uses the resource benefits both current and future generations. Sustainable solutions that improve the availability of water for human use while reducing the effects of current development on the natural system include modifications and efficiency improvements to existing water infrastructure, off-stream storage, and groundwater storage. In addition, efforts to alter storage management to allow river flow to mimic a more natural system—i.e., high spring flows—provide benefits in terms of fish habitat, channel maintenance, and water quality.

Bringing the heavily developed western rivers back to sustainable systems also requires the elimination of factors that promote use that cannot be sustained either economically or environmentally. Among the current practices considered to promote unsustainable use of water resources are the huge federal subsidies associated with irrigated agriculture. Subsidies eliminate incentive to use water efficiently and distort the true market value of the particular use. This in turn raises the cost of reallocation of water to meet other needs.

In addition to maintenance of natural capital and development of manufactured capital, sustainable use requires the development of adequate human or social capital. In the water context, this requires the development of knowledge about the water resource and the effects of development on that resource.

In addition, future generations will benefit from the establishment of institutions with sufficient flexibility to adapt water management to changes in supply and demand and to resolve future disputes over water as they arise. As recommended by Powell, these institutions must mimic basin boundaries by including participation by each jurisdiction allocating or managing water within the basin.

These features—a comprehensive solution in both participation and issues and an equitable distribution of the benefits of use of the water resource, both among members of living generations and between living generations and future generations—form the basic framework for evaluation of a durable solution. However, any solution, no matter how comprehensive and equitable on paper, will not endure if implementation cannot be accomplished with relative ease. This is the final factor in a durable solution.

Ease or Implementation

The ease with which any solution can be implemented turns on 1) the need for additional steps for finalization, 2) the simplicity and flexibility of administrative measures, 3) the likelihood of funding for implementation, and 4) the adequacy of mechanisms to resolve future disputes.

Steps to Finalization

Resolution of water disputes addressing basin-wide problems frequently requires multiple steps to achieve final approval. Particularly where governmental entities are at the table, these steps may involve substantive review and public hearings or legal proceedings. For example, if tribal water is at issue, congressional approval may be necessary. At the state level, both the Truckee and Milk River Basin agreements described in Parts I and II of this series require state court approval to assure that the agreements become part of any court decree or administration of water rights. Tribal approval often includes a referendum by the voting membership of the tribe.

Each of these steps involves considerable effort to educate a widening circle of people necessary to approve an agreement. Each of these steps can bring an agreement to a halt. However, given the need to maintain accountability to the public-at-large, it is crucial that these steps remain a part of any process. Thus, a process must be designed to increase the likelihood that these steps will proceed smoothly. Early involvement of legislative and congressional staff, of local congressional representatives and legislators, and of the public in general can facilitate final approval. A leadership role by local legislators and congressional representatives will greatly enhance both the speed and likelihood of final approval.

Finally, of greatest importance is arguably the mere recognition that these steps exist, are necessary, and will take time. Dedication by parties to a negotiation of adequate funding and staff, including staff familiar with legislative or court approval processes is essential. A process viewed as complete once an agreement is negotiated may never lead to any more than a paper solution.

Simple, Flexible Administration

Administrative simplicity and flexibility are key to easy implementation. One certainty does exist in western water—supply and demand will change. A process that results in a solution so detailed that it cannot adapt to these changed circumstances will not endure. Adaptive management is a term used to describe management that includes monitoring to detect the effects of a particular action, and a process for change in management if monitoring indicates a need for change. Adaptive management can provide the flexibility necessary to meet changing supply and demand, and the information required to know when change has occurred.

Flexibility, though important, Hill also prove awkward if methods of allocating water or monitoring compliance are too complex. Simplicity is key, particularly in the allocation of water between jurisdictions. A process that bases major allocation decisions between jurisdictions on complex models is likely to fall or become obsolete in a brief period of time. A method that requires frequent agreement between high-level representatives of, or legislative bodies of, jurisdictions prior to daily or weekly water allocation is also likely to fail. Simple measures of compliance and up-front rules on how to comply Hill endure.

Two process elements aid saddling minor adjustment to change with high level approval processes. A framework agreement generally requires approval at the level of legislatures, Congress, and the courts. By delegating the details of implementation to a separate process, such approval is not required each time an adjustment in administration is necessary. Second, joint gathering and analysis of technical data required for negotiation Hill not only facilitate the negotiation itself, but will provide the basis for formulation of implementation plans. The data gathered, the models developed, and the relationships built will facilitate implementation.

Funding

A solution must be funded to be implemented. The days of major federal spending on large-scale water infrastructure are over. Solutions must work within the limits of existing infrastructure and minimize costs to obtain funding. Local entities involved in negotiation must look to solutions that are consistent with their own budgetary constraints. Nevertheless, as noted above, enormous inequity exists in the current water infrastructure when tribal water development is compared to that of their neighbours.

The current process of funding tribal water development through the normal Bureau of Indian Affairs budgetary process merely pits that development against other tribal programs. Efforts to improve funding in this area will be discussed in the sections addressing recommendations. In addition, the enormous environmental cost of development of the rivers of the West and the cost of much needed repair to existing development are debts that must be paid by this generation if we are to leave a sustainable resource to future generations. Sections discussing recommendations point to the need for a national commitment to achieve this goal.

Future Dispute Resolution

Finally, implementation of a durable solution must include mechanisms for future dispute resolution. Unlike litigation, a negotiated solution may act prospectively to consider the likelihood that with changes in supply and demand, new conflicts will arise. An important aspect of the legacy left to future generations is a means to resolve those disputes without starting over. Thus, once again, the parties at the table must recognize that negotiation of an agreement on allocation and management of water merely begins the process of improving water supply in western water basins.

An Efficient and fair Process Leading to a Durable Solution

What appears in the preceding paragraphs is a framework for a process that is necessary to formulate an enduring solution. To summarize:

1) An efficient process achieves resolution of the dispute without excessive delay or cost.
2) A fair process provides a forum for the airing of all interests.
    a) Water use interests are invited to the table.
    b) Basin community interests are allowed comment while solutions are being formulated.
    c) The interest of the public-at-large is represented by both national standards and the appropriate level of governmental representation.
3) A durable solution:
    a) is comprehensive with respect to both parties and issues;
    b) achieves an equitable distribution of water resources benefits both within the current generation and between generations; and
    c) can be implemented with relative ease.

These abstract elements of an efficient and fair process that result in a durable solution are best understood in application to real problems. The following section takes this framework and applies it to the processes used in the Milk River Basin of Montana and the Truckee River Basin of California and Nevada.

An Efficient and fair Process and a Durable Solution

Solutions arrived at in negotiations among representatives of the United States, the State of Montana, and the Gros Ventre and Assiniboine Tribes of the Fort Belknap Reservation, pertaining to basin-wide management and allocation of water in the Milk River Basin of Montana, are discussed. Solutions arrived at in negotiations among representatives of the United States, the States of Nevada and California, the Pyramid Lake Paiute Indian Tribe, and Sierra Pacific Power Company, pertaining to basin-wide management and allocation of stored water in the Truckee River Basin of California and Nevada, are discussed. The following section analyzes those same solutions and the processes used to achieve them in the context of the framework outlined above for seeking an efficient, fair, and durable outcome.

The Milk River Basin

The Milk River in north-central Montana is a testament to our changing federal policy toward the development and use of public lands and water in the West. In 1805, when Meriwether Lewis described the milky stream laden with glacial silt entering the Missouri River, the stream's banks were home to buffalo, grizzly bear, and wolf. The Blackfeet and Sioux Indians hunted the territory of the Milk River. Today, the grizzly bear and wolf are gone and the only buffalo are in an introduced herd on the Fort Belknap Reservation. The Milk River Basin is now home to four Indian reservations and numerous Indian allotments, and it is the site of one of the earliest reclamation projects developed by the federal government.

The basin is also the locus of the dispute that led to the Winters Doctrine—the recognition by the United States Supreme Court in 1908 of Indian and federal reserved water rights. More recently in the basin history a national park and several national wildlife refuges have been established, and bull trout (Salvelinus confluentus)—a listed species under the ESA—have been found in its upper tributaries. The recognition of reserved water rights in 1908 did not settle the quantification of those rights.

The Montana Water Use Act established a general adjudication for all water rights developed under state law prior to July 1, 1973, and for all federal and Indian reserved water rights. Reserved water rights occur in every major water basin in Montana. Montana is a headwater state for the Columbia, Missouri, and Saskatchewan Rivers.

The state contains 28% federal or Tribal land, 69% of which is reserved. Of the 85 subbasins in the state identified for purposes of water rights adjudication, 70 contain claims for reserved water rights. As part of the 1979 amendments to the Montana Water Use Act, the Montana legislature established the Montana Reserved Water Rights Compact Commission (Compact Commission), to negotiate "compacts for the equitable division and apportionment of waters between the state and its people and the several Indian tribes claiming reserved water rights within the state." The Compact Commission consists of:

(a) two members of the house of representatives appointed by the speaker, each from a different political party;
(b) two members of the senate appointed by the president, each from a different political party;
(c) four members designated by the governor; and
(d) one member designated by the attorney general.

The Compact Commission acts on behalf of the state and its citizens as a whole. It represents the interests of water users as part of the general state interest, but not as individuals. To ensure that the interests of the public and individual water users are addressed, two additional steps are required by Montana law.

First, negotiated compacts must be ratified by the state legislature. Second, after ratification a compact must be filed with the Montana Water Court, which then considers the rights of individual water users claiming water in the state adjudication, and enters the negotiated water right in a final decree, integrating it with other water rights in the basin.

In 1979, when the State of Montana launched a new program for the resolution of reserved water rights through negotiation, it identified the Milk River Basin as its highest priority because the demand for water already exceeded supply and five Indian reservations claimed water rights in the basin that were not yet quantified or developed. In 1997, after years of data collection, negotiations began in earnest among the State, the Gros Ventre and Assiniboine Tribes of the Fort Belknap Reservation, and the United States. Similar to tribal water negotiations throughout the West, negotiators soon realized that no settlement would be possible without addressing issues of basin-wide concern.

Efficiency

The agreement on the Milk River arose out of efforts to quantify the reserved water rights of the Gros Ventre and Assiniboine Tribes of the Fort Belknap Reservation. The Reservation and the basin upstream of it was the location of the water dispute that In 1908 led to the recognition of reserved water rights by the United States Supreme Court. The Reservation is located at approximately the mid-point on a river that includes the numerous jurisdictions. Delivery of water to Fort Belknap requires coordination among these entities.

Diversions upstream can prevent that delivery. Yet in 1997, ninety years after Winters, when negotiations began in earnest to resolve the Tribes' water rights, no comprehensive tool existed for coordinated management of water in the Milk River Basin. The agreement negotiated between 1997 and 2001 will probably take at least an additional five years to achieve final approval. Thus, the dispute has taken roughly one hundred years to settle. This is not efficient. However, when broken down into years spent seeking solutions in court and years spent In negotiation, it is clear that the negotiation effort moved at a rapid pace.

The State of Montana established a process for negotiated resolution of tribal water rights in 1979. Between 1979 and 1995, the parties to the Fort Belknap Compact invested in joint development of a hydrologic model of the Milk River, a Geographic Information System (GIS) database of irrigation in the Milk River Basin, and joint review of public information on arability of soils on the Fort Belknap Reservation and irrigation within the Reclamation project.

That joint data-gathering effort led to the rapid settlement of water rights of the Fort Belknap Reservation in the period between 1997 and 2001, including use of the GIS database on irrigation as a basin-wide management and enforcement tool. Furthermore, the three year negotiation process not only led to quantification of the water rights of the Tribes, but to a tool for comprehensive management and enforcement to ensure delivery of that right. Through negotiation the parties were able to avoid a piecemeal approach to solving problems across jurisdictional boundaries.

Fairness

The fairness element turns on the inclusion of a widening circle of interests from water interests, to basin community interests, to the public-at-large. Each interest is viewed separately for the Milk River negotiations.

Water Interests

The parties at the table in the Milk River negotiations were the United States, representing the interests of the Tribes, the U.S. Bureau of Reclamation, and the U.S. Fish and Wildlife Service; the State of Montana, representing the interests of the State and of the people claiming water rights pursuant to state law; and the Fort Belknap Reservation. The irrigation districts were absent from the table. Under Montana law, their interests and those of other off-Reservation water users in the basin were represented by the State of Montana.

Montana takes the position that negotiations with tribes must be on a government-to-government basis. To ensure that the water right claimants in the basin have a voice at the table, the Commission representing Montana provides for extensive public involvement, including individual meetings with irrigators and irrigation districts. However, the Commission also takes the position that if a conflict arises between general state interests and the interests of individual water users, the state interests will prevail.

Examples of other state interests include: interests of the Department of Fish, Wildlife, and Parks in fisheries and instream flow; interest in resolving the water conflict; and interest in keeping the settlement's impact on the state treasury to a minimum. To give voice to individual water users in light of the fact that their concerns may be subordinated to other state concerns in negotiations, water right claimants have the opportunity to object to entry of the agreement in a decree by the Montana Water Court. The court has the choice of either approving or voiding the agreement. Although this step ensures that no agreement will become final if it infringes on the rights of water right claimants, it does nothing to obtain their proactive participation in developing solutions. In Montana, that is left to the voluntary participation of water users in the public participation process established by the Commission.

In addition to state efforts to involve the public, during the negotiations the representatives of the U.S. Bureau of Reclamation met extensively with irrigation districts. However, the role of the federal team in negotiations is as trustee for the tribes. Thus, although the Bureau of Reclamation provided substantial technical information concerning the federal project, it was not participating in negotiations as a representative of the interests of the project water users. Filtering of the participation of local water interests through state representatives may prevent full consideration of local solutions. Of equal importance, absence of representatives of irrigation districts from the table may hinder implementation.

Local buy-in by those who will live with the results of a negotiation on a daily basis is essential to the full realization of its benefits. In addition, by allowing private interests to participate at the table, state and federal governments can represent the general public interest with less internal conflict over local interests. Although the State's respect for the governmental status of the Tribes is important, respect for tribal status has not prevented the inclusion of important private interests in negotiations in the Truckee River Basin. Though certainly an important consideration, the interests of the State and Tribe in a fair and durable agreement weigh in favour of the presence of private parties at the table.

Basin Community Interests

Interests of the wider basin community surface at two levels in the Milk River Basin negotiations. First, the extensive public involvement of the Compact Commission includes numerous public meetings within the basin to obtain comment from local interests in general in addition to water interests. This process was also undertaken by the Tribes on the Fort Belknap Reservation to obtain comment from the Reservation community. Both efforts had varying degrees of success in obtaining attendance at meetings. Representatives in negotiations had greater success when they asked to be included on the agenda of scheduled local meetings such as those of local Conservation Districts.

Public comment did result in numerous substantive changes to the agreement affecting water allocation and solutions to management and enforcement problems to meet the concerns of the local community. The federal representatives to negotiations undertook no effort at public involvement other than the meetings with the irrigation districts. However, federal representatives attended and were available for questions at most Compact Commission public meetings. Second, each governmental entity requires a final approval process that includes either representative or direct vote. The State of Montana requires legislative approval of any reserved water right settlement. Legislative hearings on the Fort Belknap compact included testimony from irrigation districts. Congressional and tribal approval of the Fort Belknap Compact is pending. Tribal approval will include a referendum vote by the eligible membership of the Fort Belknap Indian Community.

Both the public involvement and the finalization process in the Milk River Basin assure ample opportunity for comment by basin community interests. By including an early phase of public involvement, the process in the Milk River negotiations ensures that comments will be received during the formulation of solutions when it can have an effect on the outcome.

Public Accountability

In the Montana effort, as in many negotiation processes, public accountability is the weak link. However, unlike most other processes, the use of a politically appointed commission provides strong accountability at the state level that can be used as a model for other processes and participation by other levels of government. The presence of state legislators from across the state on the Compact Commission and the political ties of the remaining members not only ease approval of compacts before the Montana Legislature, but also provide a degree of accountability to the general public. This accountability is weakened by the fact that the Commission takes the narrow view that its primary role is to protect existing water use. To remedy this weakness, the legislature could direct the Commission to consider a broader perspective including the health of Montana's watersheds and economic development for its citizens, whether on or off reservations. Leadership would then be necessary to balance these broader interests against the vocal interests of local water users.

The general public represented by the Compact Commission stops at the state line. The role of the Compact Commission as a state entity does not include giving voice to concerns of the national public. The primary means of accountability to the national public are the requirements that any project negotiated comply with national environmental laws including the National Environmental Policy Act (NEPA) and the ESA, and receive congressional approval.

The NEPA process provides a procedure that allows public comment, but, because it occurs after a project has been authorized by Congress, it is unlikely to alter the choice of a negotiated solution. To give greater weight to environmental standards in influencing the choice of alternatives, the Fort Belknap Compact avoids choosing specific alternatives for improvements to the reclamation project contemplated as part of the agreement. Instead, the Compact delegates the choice to the Bureau of Reclamation after environmental review.

In addition, congressional approval should in theory provide accountability to the national public. However, the structure of congressional committees involved in water issues generally provides a narrow focus in which economic interests in existing water development, particularly federal projects, prevail. Furthermore, consideration of a larger public interest at the late stage of congressional approval, after interests in negotiated solutions have firmly developed, is arguably too late to influence the details of the agreement. Whether these weaknesses in representation of the public-at-large in designing solutions for the Milk River Basin hindered the durability of the outcome.

Durability of the Outcome

Evaluation of the durability of the Fort Belknap Compact requires consideration of the comprehensiveness of the solution, whether the solution results in an equitable distribution of water-use benefits both within and between generations, and the ease with which the agreement will be implemented.

Comprehensiveness : Comprehensiveness of a solution refers to both the parties represented at the table and the issues addressed. One major player in the United States's portion of the Milk River Basin was not at the table in the negotiations to settle the water rights of the Fort Belknap Reservation: the Blackfeet Tribe. The Blackfeet Reservation is located in the headwaters of the Milk River and is the location of a major interbasin water diversion that serves the federal reclamation project.

However, the Milk River process contemplates a phased approach to settlement of all issues in the basin. The Blackfeet, the United States, and the State of Montana continue to negotiate. In addition, the Fort Belknap Compact delegates study and design of improvements in water infrastructure to the United States Bureau of Reclamation. Any decision to include portions of the upper basin in the improvements cannot be made without an invitation to the Blackfeet Tribe sit at the table. This phased approach recognizes the overwhelming complexity of including every interest in a basin in a single agreement. Attempts in the early 1990s to develop a comprehensive Milk River Basin tribal water settlement failed due to the complexity of relations between parties and the issues. Once again, it remains to be seen if the phased approach will achieve a comprehensive settlement in the end. Issues addressed in the Fort Belknap Compact are limited to those involving water allocation and use. The agreement addresses both surface and groundwater, but does not address water quality. This lack of connection between water right quantification and water quality reflects the basic separation between the laws governing the two areas in the western United States. However, it is encouraging that the loan program established by the agreement allows consideration of improvements to water quality by the Milk River Coordinating Committee in issuing loans for projects to improve irrigation efficiency. In addition, a basin-wide coordinating committee, established by the agreement to coordinate management of water between jurisdictions creates an institution that the basin could use to address water quality issues in the future.

Equitable Distribution of the Benefits : Within the current generation, the Milk River agreement, if ultimately approved and funded, will go a long way to remedy the inequitable distribution of federal water development dollars between the Fort Belknap tribes and their neighbours. Federal funds spent in the twentieth century were allocated predominately to off-reservation water development. Parties to the agreement are currently negotiating the federal bill for approval of the agreement. Included in that approval will be authorization of funding for development of tribal water. At the same time, the agreement maintains the existing dependence of the rural economy on irrigated agriculture by authorizing study and design to improve the current reclamation project.

Any significant improvement or renovation to the project will require a substantial federal subsidy in a basin where making a living on irrigated alfalfa and pasture is marginal at best. Under the standards proposed by the Western Water Policy Review Advisory Commission, such continued reliance on federal subsidy does not result in sustainable use of the water resource. The dilemma facing the citizens of the Milk River Basin is that once a community has gone down the path of economic reliance on a federal irrigation project, it is difficult to turn back. Yet continuing down that path requires continued infusion of federal funds. Studies are currently underway by the United States Bureau of Reclamation to determine exactly what changes to existing water infrastructure in the Milk River Project are appropriate.

Unfortunately, without major changes that include broader consideration of a sustainable economic base, future generations will likely face similar problems to those currently before the people of the Milk River Basin. However, it is encouraging that the Fort Belknap Compact established a loan program for improvements in water use efficiency and a drought water bank that can shift water, on a voluntary basis, to critical needs including habitat. In addition, indications are that the final solutions for improvement in water infrastructure will most likely work within the confines of the existing system, rather than attempt to develop major new water sources.

Although it is in part due to the investment of previous generations in water infrastructure that Americans have cheap power and do not stand in lines to purchase food, this policy basis for irrigation subsidy is weak when applied to the question of whether there are better uses for public money than subsidizing alfalfa in an arid region with a northern climate.

Nevertheless, it is national policy, followed in the early 1900s, that placed the Milk River Basin in its current dilemma. Sustainability requires that new projects move toward the elimination of subsidy. However, the Milk River process indicates that locally driven negotiations are unlikely to eliminate subsidies in basins whose economic bases are dependent upon such subsidies. In a basin with a history that is not one of large profits by agribusiness, but rather one dominated by individual family farms (the goal of twentieth century reclamation policy), this may also reflect a sense that it is inequitable to place the entire burden of past mistakes on the people who have come to rely on irrigated agriculture.

Finally, it should be noted that the decision was made almost a century ago to transform the Milk River from a river into an irrigation delivery system, to varying degrees, for the reservation and nonreservation lands. Without the interbasin transfer of water, the Milk River did not run during dry summer years. No issues concerning fisheries have yet arisen in negotiations on the portion of the river affected by the Fort Belknap Compact.

The long use of the Milk River as an irrigation conveyance system and its minor contribution, at its terminus, to the much larger Missouri River, render identification of endangered species reliant on Milk River water flow unlikely. It is possible that studies to identify appropriate improvements to the Milk River Project may discover environmental issues. More likely, it will come down to a choice of infrastructure improvements reflecting current values within the basin. Until a national voice on the future of the federal project is heard, those values, heavily influenced by local interests, are likely to maintain the Milk as an irrigation delivery system.

Ease of Implementation : The ease with which an agreement can be implemented turns on the complexity of the steps necessary to finalize the agreement, the simplicity and flexibility of its administration (particularly between jurisdictions), the availability of funding, and mechanisms for future dispute resolution. First, the Fort Belknap Compact received approval by the Montana Legislature in 2001. This process was quick and easy due to the presence of legislators on the Commission. Remaining steps to finalization include congressional approval, entry of a decree by the Montana Water Court, and a Tribal referendum.

Although each of these steps is difficult to achieve, the need for accountability to the broader public warrants the effort. To avoid surprises in the final approval stages, the parties to the Fort Belknap agreement included public involvement and frequent briefing of legislative and congressional staff as part of the process. Nevertheless, lack of active participation by congressional staff or Congressmen may hinder that process. Second, the Fort Belknap Compact establishes a basin-wide coordinating committee to coordinate water storage and management among tribal, state, federal, and irrigation district jurisdictions. Even more innovative is the agreement to jointly appoint water commissioners to enforce the daily distribution of water from the mainstem of the Milk River.

In a complex system with interbasin transfer of water, stored water, and multiple jurisdictions distributing water, this simple agreement to seek a single enforcement entity should go a long way toward easing concerns over implementation among the parties. More complex is the accounting system the United States Bureau of Reclamation must develop and the database water distributors must maintain to determine when to release water and where it must go. Water flowing in the bed of the Milk River includes water imported from the St. Mary River, stored water, and the natural flow of the Milk River. Separating out these sources and who has claim to them on a daily basis is a daunting task. However, by delegating the tasks of establishing an accounting system and database to actions separate from the actual agreement, and allowing any system established to be modified on the basis of new information, the agreement allows flexible management that should prove helpful to meet changing circumstances and variable water supply.

Third, both the size and sources of funding for implementation of the Fort Belknap Compact are currently under negotiation as part of efforts to develop a federal bill for ratification of the agreement. The likelihood of receiving funding will turn partially on whether the amount sought is within state and federal budgetary limits, and partially on whether Congress removes existing barriers to funding of Indian water right settlements. The problems with the current funding process and the efforts to change it are discussed in subsequent sections. Finally, the Fort Belknap Compact establishes a process for resolution of future disputes. The agreement sets up a Compact Board composed of a state and tribal representative and a third member chosen by the other two. The Compact Board is authorized to hear disputes over interpretation of the agreement and any disputes between an individual authorized to use water under state law and an individual authorized to use water under tribal law.

By setting up a joint state-tribal board, the Compact gives voice to both governments with jurisdiction over water use in the basin, ensuring equal participation in the resolution of future disputes. The process followed in the Milk River differed strikingly from that employed on the Truckee River. Thus, analysis of the Truckee River will provide a perspective on results using a different approach. In addition, unlike the Milk River, urban needs, fisheries, and interstate apportionment dominated discussions on the Truckee River, allowing analysis of a different set of problems.

The Truckee River Basin

The Truckee River takes its water supply from the snowpack of the Sierra Nevada Mountains in California, where five major federal and several private reservoirs regulate its flow. The terminus of the Truckee River, Pyramid Lake, is located within the Pyramid Lake Paiute Indian Reservation. Years of litigation attempting to reallocate water to Pyramid Lake upheld the dominance of appropriative water rights for the federal reclamation project that diverts water to the Carson River Basin known as the Newlands Project. Diversions of the river to satisfy the irrigation project resulted in lower lake levels, which blocked fish passage to spawning grounds. The Pyramid Lake cutthroat trout disappeared entirely from the lake in the late 1930s or early 1940s, though a similar strain of Lahontan cutthroat trout was introduced subsequently. The Lahontan cutthroat trout was listed as threatened under the ESA in 1975. The cui-ui, another important fishery dependent on the lake and upstream spawning habitat, was listed as endangered in 1967. Meanwhile, the urban areas of Reno and Sparks in Nevada grew, placing an increasing demand on water for municipal needs, particularly during drought. Finally, despite years of negotiation, California and Nevada had never reached an apportionment of Truckee River and Lake Tahoe water that could achieve congressional approval.

In 1990, after years of litigation and less than comprehensive negotiated agreements, Congress passed the Truckee-Carson-Pyramid Lake Water Rights Settlement Act (1990 Settlement Act). Among other things, the Act required a process to revise the operating criteria for the Truckee River for the restoration of endangered species and to provide a drought water supply for urban areas, as well as authorizing changes to the operation of federal dams for these purposes.

The modern trend toward altering river management to meet changing needs is embodied in the 1990 Settlement Act, and the effort to implement that Act through the pending Truckee River Operating Agreement (TROA) If finalized and adopted, implementation of the TROA is likely to aid in the restoration of the Pyramid Lake fishery, ensure a drought water supply for Reno and Sparks, and achieve an apportionment between California and Nevada.

A long and contentions history brought the parties to the brink on which the 1990 Settlement Act was negotiated. Certain elements in this history stand out as key in setting the stage for comprehensive resolution of many of the basin's water disputes. They can best be understood in the context of the parties they brought to the table:

1) California. Efforts in the 1970s to achieve congressional approval of an Interstate Compact between California and Nevada regarding the Truckee, Carson, and Walker Rivers failed, largely due to objections by the Tribe and the United States. As a result of that failure, water users in the Truckee River Basin in California faced a serious threat of loss of the ability to use water to the senior fishery flow water rights claimed In a suit filed by the Tribe. In addition, due to lack of an apportionment, the California State Water Resources Control Board has been unwilling to issue permits for new water use since the 1970s. California considered certainty on apportionment and how tribal water rights would be treated in that apportionment to be key to providing a foundation on which operation of the Truckee River Basin could occur.

2) Nevada Nevada, like California, was frustrated with failure of the interstate compact. As a result of that failure, Nevada's water users in the Truckee River Basin faced the constant insecurity posed by upstream diversions. Furthermore, almost all storage on which Nevada relies is located in California. Nevada considered interstate apportionment to be its primary goal in entering negotiations on the 1990 Settlement Act, but also sought to resolve major concerns regarding a drought water supply for the Reno/Sparks area and settlement of litigation concerning the Newlands Project.
3) Pyramid Lake Paiute Indian Tribe. The United States Supreme Court foreclosed the Tribe in its efforts to argue for tribal water rights for fisheries. But the Tribe's success on other fronts turned the tables, allowing it to bring a strong voice to the table. First, the Tribe was in the fortunate position of having its interest in fishery flows aligned with the newfound national interest embodied in the ESA. Second, the Tribe successfully asserted the fiduciary duty of the United States and, although foreclosed on the issue of reserved water rights for fisheries in Nevada, brought the federal fiduciary duty to bear on the operation, management, and efficiency of water use in the Newlands Project. Finally, by asserting new reserved rights for fisheries against California, the Tribe got the attention of the upstream state and guaranteed that the powerful California congressional delegation would support any settlement.
4) Sierra Pacific Power Company (Sierra Pacific). Sierra Pacific's urban water users in the Reno/Sparks area of Nevada had fallen last in line, since they had outgrown their allocation in the 1944 Orr Ditch Decree and were being served through purchase of irrigation water rights beginning in the 1940s. Sierra Pacific needed a firm supply of water during drought. What Sierra Pacific brought to the table was, in part, the high economic value of urban water. In addition, Sierra Pacific was the beneficiary of instream flow requirements set in 1908 to assure water for small hydropower stations and referred to as the Floriston rates. Their willingness to waive those rates, which did not mimic the natural river fluctuations necessary for habitat, became key to successful negotiations with the Tribe.

5) Truckee Carson Irrigation DistriCt (TCID). TCID represents the water users in the federal Newlands Project. TCID initially participated in the 1990 Settlement Act negotiations, but withdrew periodically when it believed that its interests were not being addressed, and is not participating in current TROA negotiations. The State of Nevada and the United States have attempted to represent the interests of TCID in current negotiations, but believe the first challenge to the final TROA will come from TCID.
6) The United States. Representatives of the United States Departments of the Interior and Justice have participated throughout negotiation of the 1990 Settlement Act and TROA, taking a lead role in convening TROA negotiations. The federal interest in the Truckee River Basin has been strong throughout its modern history and includes the needs of the Pyramid Lake Paiute Indian Tribe as trustee, the integrity of the federal reclamation project, the federal reservoirs, and the endangered species, specifically the cui-ui and the Lahontan cutthroat trout.

One additional element for successful negotiation of the 1990 Settlement Act was leadership from Senator Harry Reid of Nevada (D-Nev.) who used the power of his office to facilitate, supervise, and push negotiations to a final result, Under Senator Reid's leadership, parties to the final agreement were assigned the task of entering separate negotiations focused on specific issues.

The goal was to allow resolution of issues in manageable bites, then to assemble each agreement into a comprehensive whole. The result was as follows: the Preliminary Settlement Agreement (PSA) negotiated between Sierra Pacific and the Tribe covering issues of fishery flows and urban water supply during drought was completed in 1989; the 1990 Settlement Act resolving the interstate issues and incorporating the PSA; and the TROA, authorized by the 1990 Settlement Act to cover the operation of the upper basin reservoirs, which is currently under negotiation.

Efficiency

Roughly a century of litigation and failed attempts at settlement set the stage for the current process to negotiate a solution to water management, interstate apportionment of water, and restoration of fisheries in the Truckee River Basin. A plan to accomplish these objectives has taken approximately 15 years to construct. In water disputes as complex as those faced in the Truckee River Basin, fifteen years is an efficient use of time.

However, the preceding 85 years of litigation cannot be discounted. Without the resolution of certain issues through litigation, the shift in power resulting from successful litigation to improve water flow to Pyramid Lake, and frustration over the inability of litigation to provide a satisfactory solution, the current settlement may not have been possible. Once the stage was set, negotiation provided an efficient means to achieve much broader solutions than would otherwise have been available.

Current negotiations in the Truckee River Basin are focused on the renegotiation of the 1998 TROA. Parties to the negotiation indicate that the 1998 TROA became unacceptable when Sierra Pacific realized that its modelling assumptions used to predict water availability for its customers were not consistent with the lake levels and tributary instream flow maintained in the upper basin under the 1998 TROA. This inconsistency illustrates the inefficiency caused by independent use of a hydrologic model by individual parties.

The model in this case was developed by the Bureau of Reclamation, but modified by Sierra Pacific during negotiations of the PSA between the Tribe and Sierra Pacific in which the focus was on urban and fishery water needs in the lower basin. Additional parties to the TROA negotiations—the United States and California—expanded the scope of negotiations to include upper basin interests in lake levels for recreation and instream flow below dams on tributaries for Lahontan cutthroat trout and other fisheries, yet these interests were not reflected in the model being run independently by Sierra Pacific.

Establishment of a joint team consisting of technical representatives from each party would avoid the inefficiency of differing assumptions in modelling solutions. The chair of the federal team on the Truckee River negotiations recommends hiring a neutral consultant and full documentation of the model and model runs.

Although this approach would alleviate the problem caused by working from different assumptions, it may not improve efficiency. Each party to the agreement would still assign separate technical representatives to review the neutral model with the potential for differing results. Efficient use of time in negotiations requires focus on substantive issues. By allowing technical representatives from each party to work together to develop a model, disagreement over model assumptions, use, and interpretation of results would be minimized.

Fairness

Similar to the analysis of the Milk River process, the fairness element turns on the inclusion of a widening circle of involvement from water interests, to basin community interests, to the public-at-large. Negotiations in the Truckee River Basin approach this through a very different process than that used in the Milk River. Those differences are noted in the following sections.

Water Interests : Unlike negotiations in the Milk River Basin, both public and private entities are at the table in the Truckee River negotiations. Participation appears to turn on the level of stake in water use and storage management rather than governmental status. Thus, the initial players included California, Nevada, the Pyramid Lake Paiute Indian Tribe, Sierra Pacific Power Company, and the TCID. As noted above, allowing key water interests a seat at the table provides a greater possibility that basin-specific solutions will be recognized, water interests will buy-in to solutions, and the ultimate agreement will be implemented.

Unfortunately for the Truckee River parties, a key player—TCID-chose to withdraw from the process. This withdrawal diminishes the comprehensiveness of the agreement. The withdrawal may not diminish the fairness of the process because it was voluntary. However, it is important to note that the stated reason for withdrawal was failure by the other parties to fully honor TCID's view of the scope of its valid water rights in design of solutions.

If TCID's perceived unfair treatment is real, then similar to the process in the Milk River the ability to challenge the agreement in court provides a final safety net, albeit one detrimental to the durability of the agreement.

Basin Community Interests : Formal local public comment is left to the final approval stages of negotiation in the Truckee River Basin. On completion of the Truckee River Operating Agreement, public comment will be received in an environmental review process required by both the United States and the State of California. In addition, the congressional process that led to the 1990 Settlement Act provided an opportunity for public comment. Final approval of the TROA will require a political process by each governmental entity. This process will include a tribal referendum vote.

Unlike the Milk River process, the Truckee River process will obtain the views of the local community after solutions have been designed and have received preliminary approval by representatives of the parties. This approach tends to miss additional opportunities for local design of solutions and to render public comment a mere procedural step.

The presence at the table of private water managers representing key urban and agricultural interests in the basin does provide substantial opportunity for local input. Nevertheless, instream interests other than the fisheries represented by the Tribe and the United States are left to representation by state participants. This approach may prove inadequate.

Many negotiations are closed to maintain confidentiality on data and positions prepared for litigation. In addition, parties to a settlement may be understandably nervous about public and media scrutiny of options under early stages of discussion. Furthermore, such scrutiny may have a chilling effect on open discussion of ideas. However, a careful control of the release of information would address these concerns while serving the need to inform the public. First, by maintaining confidentiality on technical information and parties' positions and by releasing only information on solutions under consideration, parties can protect the information generated for litigation. Second, by identifying solutions not yet studied as premature for release, parties can avoid speculation on preliminary ideas, ensuring that possible solutions are released to the public when the parties have sufficient information to explain their effects. Third, avoiding attribution of a settlement option to any particular party can eliminate the chilling effect on open discussion of ideas.

Public Accountability : Similar to the Milk River process, accountability to the wider circle of the public-at-large is generally limited to the approval stages of the various Truckee River agreements, including entry of the agreement in the Orr Ditch proceedings in the Truckee River Basin. However, Senator Henry Reid filled the role of political leadership played by the Montana Reserved Water Rights Compact Commission in the Milk River negotiations. Unlike the state Compact Commission, Senator Reid held a national political office, placing him in a better position to represent national interests. Interviews with the parties indicate that he did so by placing a high premium on achieving settlement, and also by assuring that the voice of each party was heard.

In addition, a national standard in the form of the ESA played a substantial role in shifting power among parties and in giving fishery interests a voice at the table. In doing so, the negotiations reflected the shift in national interest toward environmental values. Finally, the federal role in negotiations on the Truckee River Basin was substantially different from that on the Milk River. Whereas the federal team on the Milk River primarily represented the interest of the tribes, the federal representatives in the Truckee River negotiations took a broader view of their role, representing all federal interests reflected in federal law. Thus, the federal team represented the tribe, the federal project, and the interests in protection of endangered species.

This broader view of the federal interest allowed representation of the diverse national interests reflected in federal law and provided a greater level of public accountability than that reflected in the current federal process for participation in negotiation of tribal water rights. The difficulty of representing these potentially conflicting interests cannot be overstated. However, the participation of federal negotiators, whose position in the Department of the Interior enabled rapid decision making in the Truckee River Basin negotiations, helped ensure that conflicts among federal interests could be resolved on a timely basis.

Durability of Outcome

Evaluation of the durability of the Truckee River agreements requires consideration of the comprehensiveness of the solution, whether the solution results in an equitable distribution of water use benefits both within and between generations, and the ease with which the agreement will be implemented.

Comprehensiveness : TCID's absence from settlement in the Truckee River Basin prevents the parties from achieving a comprehensive solution as to both parties and issues. It is both a strength and a weakness of negotiation, as opposed to litigation, that participation is voluntary. No means exists to force a party to the table. One cure may be Montana's approach, where the legislature designated the state as the representative of all private interests, accompanied by a state policy in favour of negotiation. A private party's disgruntlement cannot force the state to walk. This approach assures a comprehensive solution. However, as noted above, it may not assure a solution that is readily implemented if additional efforts are not taken to obtain buy-in by the local interests.

The Truckee River Basin process takes a phased approach to issue settlement. Under the leadership of Senator Reid, issues were divided into manageable segments including allocation of surplus storage between fisheries and municipal interests reflected in the negotiation of the PSA, interstate apportionment of water, and resolution of issues involving the Newlands Project. Resolutions of the storage and apportionment issues were then approved in the 1990 Settlement Act. Finally, that Act authorized, and even required, negotiation of the much more detailed TROA to validate the other agreements.

This phased approach does not detract from the comprehensiveness of the solution; instead, it recognizes that the complexity of water management and allocation is best addressed in manageable steps. More importantly, it recognizes that while the framework for a basin-wide agreement may appropriately be subjected to congressional approval, the details of an operating agreement require a greater degree of flexibility and thus a less onerous process to adapt to changing circumstances.

Unfortunately, the absence of TCID from negotiations prevented settlement of issues related to the Newlands Project. The absence of agreement related to the project will likely require additional efforts to achieve a comprehensive solution. Because of the major role of the project in water use from the Truckee River Basin, there is no guarantee that those additional efforts will not force the parties to revisit issues currently considered settled. However, efforts by the United States and Nevada to protect the water rights for the federal project may allow them to maintain the integrity of the agreement.

Equitable Distribution of the Benefits : The end result of efforts to resolve the water issues in the Truckee River Basin is a solution that more equitably distributes the water-use benefits among the members of this generation and leaves a more sustainable resource to future generations by mitigating past environmental harm, by providing an urban drought water supply, and by establishing a joint mechanism for dispute resolution. This attention to repair of environmental harm distinguished the Truckee River settlement from the Milk River compact, and, at least at this stage, provides greater intergenerational equity. However, it cannot be said that this redistribution of benefits occurred solely as the result of negotiations.

Negotiation provided a sufficiently flexible and political forum to achieve some of the management results in the Truckee River Basin.

This was reflected in the Preliminary Settlement Agreement, the 1990 Settlement Act, and the draft TROA. Litigation, national legislation in the form of the ESA, and a shift in the economic, and therefore political, focus in the basin from agriculture to urban needs provided the shift in power necessary to set the stage for settlement.

This is an important lesson in the need for a judicial and political backdrop within which negotiation can occur. Even in the Milk River where the negotiation process dominated, without the 1908 recognition of the reserved water rights of the Tribes by the United States Supreme Court, it is unlikely the parties would have even begun the process and even less likely they would have focused on developing tribal water rights.

Thus, if the Truckee and Milk River processes are representative, the second prong of a durable outcome—an equitable distribution of benefits—appears to require both the development of national requirements and at least the threat of litigation before negotiation can provide a solution. It is not surprising that the actual reallocation of water, whether to tribes, environmental needs, or urban areas, is the step that requires the biggest stick.

Ease of Implementation : The ease with which an agreement can be implemented turns on several things including: the complexity of the steps necessary to finalize the agreement, the simplicity and flexibility of administration particularly between jurisdictions, the likelihood of funding, and mechanisms for future dispute resolution. First, the phased approach to achieving resolution of water issues in the Truckee River Basin has made finalization of agreements manageable. As discussed in evaluating the Milk River negotiations, the requirements for congressional, legislative, and tribal referendum approval, though onerous, also serve the need of public accountability. Participation by Senator Harry Reid early in the process rendered the congressional step in the approval process more likely.

Second, because the final TROA is not yet public, it is difficult to evaluate whether the solutions incorporate sufficient flexibility and simplicity to ease implementation. If the TROA is too detailed, it may lack the flexibility necessary to allow management to adjust to future changes in water supply and demand. Nevertheless, the decision to follow the existing agreement between Nevada and California on apportionment, one with a track record of compliance, simplifies implementation. In addition, the decision to use the office of the federal Water Master, already assigned the task of implementing the

Orr Ditch Decree for administration of decreed water rights, means that the task of daily administration will also be performed by an entity with a proven track record in the basin. Depending on the outcome of the TROA negotiations, these decisions bode well for the basin.

Third, funding is also uncertain until the TROA negotiations are completed and made public. However, at least the assignment of administration to the federal Water Master places them in the hands of an office that already has the authority to assess fees against water users. Thus, the funding mechanism for that portion of the settlement is built into the choice of the entity for the task.

Finally, by use of the office of the federal Water Master for daily administration and by setting up a process for appointment of a hearing officer for dispute resolution, the Truckee River agreements allow for adjustment in the face of a continually changing water resource. In doing so, the parties have set the stage for a durable agreement in the Truckee River Basin.

A Process for Basin-wide Water Dispute Resolution

The resolution of water disputes in the West relies increasingly on negotiation as people endeavor to meet changing and growing needs for water, to remedy past harm to the environment caused by water development, and to remedy the inequitable distribution of federal dollars to provide benefits from that development. Negotiation is proving successful in allowing parties to avoid the narrow single-issue approach of litigation and to fend more creative solutions to chronic problems with water management and allocation. Negotiation allows parties to recognize the shared nature of a water resource and to step beyond the jurisdictional boundaries dividing basins to find shared solutions.

Four concerns arise with the current ad hoc approach to solving major basin-wide water disputes through locally driven negotiation processes including the: 1) lack of a hammer to force parties to the table and to force consideration of difficult issues, 2) lack of accountability to the public-at-large, 3) lack of certainty that the results will lead to an equitable distribution of water-use benefits, and 4) inability or lack of will to fund solutions, particularly where water development to serve those currently receiving fewer benefits from the water resource is needed. These concerns and recommend elements for incorporation into a negotiation process to assure they are addressed.

The Hammer

The shift in power resulting from prior litigation and the threat of future litigation played a major role in bringing parties to the table and in forcing consideration of environmental and tribal issues in both the Truckee and Milk River Basin negotiations. In addition, enforceable national standards in the ESA forced consideration of habitat issues in the Truckee River Basin. Even with these incentives, TCID, a key party in the Truckee River negotiations, chose not to participate in negotiations.

The Milk and Truckee River negotiations illustrate the continued need to negotiate against the backdrop of possible litigation. A dispute resolution process that allows the freedom to negotiate while maintaining judicial oversight serves two functions. First, it allows intervention by a neutral decision maker when an impasse is met on a particular issue in negotiation or when a necessary party is reluctant to participate. Second, it facilitates finalization of any agreement that must be integrated into a water rights decree for the basin at issue. Of the negotiation processes currently underway on major water basins in the West, those proceeding in Arizona give the strongest role to the judiciary.

Under the rules of the Arizona Supreme Court, the parties to water negotiations arising in the context of a general stream adjudication may seek interlocutory review of legal issues that require resolution for the parties to reach agreement. However, it is important that parties not use the availability of a judicial forum as a crutch to avoid addressing difficult issues or to unnecessarily constrain negotiation. The Arizona Supreme Court imposed highly limiting constraints on tribal water right negotiations by ruling that the amount of water allocated in any settlement may not exceed the amount that would have been available in court. This has caused parties in Arizona to spend considerable time debating the legal measure of tribal water rights rather than seeking creative solutions to water supply problems.

Often, it is the uncertainty in the law—i.e., the fact that a party may not know what the outcome in court will be but knows they are at risk—that leads parties to seek creative solutions. Unnecessary constraints on negotiation could be avoided by resorting to a judicial forum primarily for procedural rather than substantive issues to ensure, for example, that all parties are present, have notice of proceedings, and adhere to a timeline. A high bar to litigation on substantive issues should be imposed both through procedural mechanisms that require prior efforts to negotiate and through judicial constraint.

In addition to the threat of litigation to allocate water rights, enforcement of national standards reflecting changes in national values with regard to water use provide a tool for encouraging change. Of particular importance in the Truckee River Basin were the requirements under the ESA that allowed dedication of surplus federal project water to fishery flows. Continued threat of enforcement of laws to protect endangered species and water quality is necessary to force consideration of issues of national concern that are not otherwise represented at the table in a local water basin dispute. The use of national standards for this purpose leads directly to the issue of accountability to the public-at-large.

Public Accountability

Maintaining accountability to the public-at-large is the weakest link in current processes to resolve water disputes in the West. As is evident in both the Milk and Truckee River processes, negotiation opens the door to considerable local innovation but does not provide an avenue to secure representation of the interests of those absent from the table. The public interest includes the health of the federal lands, the expenditure of federal tax dollars on water infrastructure, and the long-term sustainability of the use of the nation's water resources.

The current federal process for participation in water negotiations represents the proprietary interest in public and reserved lands and the interest in prudent use of federal tax dollars. However, the sustainable use of water resources is not a mandatory element of a negotiated solution unless a listed endangered aquatic species is at risk. Incorporating the need for sustainability into water dispute resolution processes requires the following: 1) national standards that basin-wide negotiations must meet, 2) reform of the current federal process for participation in tribal water right negotiations and formation of a federal process for participation in water negotiations in general, and 3) strong political leadership from both Congress and local legislators.

National Standards: National standards that dictate the format or content of a basin-wide water management or allocation solution would have a chilling effect on the creative solutions currently blossoming under the freedom of negotiation. However, focused standards in areas of national concern that might otherwise be ignored are important to assuring a national voice. For example, the twin hammers of the ESA and the Clean Water Act provide an incentive to negotiate and a requirement to accommodate issues relating to the long-term viability of the resource and downstream effects of its use. With the need to develop tribal water and to modify or renovate existing federal projects such as the Milk River Project, the time is ripe to discuss federal priorities for funding of renovation and allocation of water in federal water projects in the West. This issue will be discussed in the context of the equitable distribution of benefits and funding for water development. For purposes of public accountability, the outcome of any national prioritization on funding and use of federal projects will sufficiently represent the interests of the public-at-large.

Reform of the Federal Process : Because accountability of the process of water dispute resolution through negotiation is a concern to the public, federal participation in negotiation is a natural starting point to address that concern. However, the same principles apply to state and local government in designing their own processes to assure public accountability. The current federal process for participation in water negotiations in any basin involving claims to tribal water is set forth in the Criteria and Procedures for the Participation of the Federal Government in Negotiations for the Settlement of Indian Water Right Claims (Criteria and Procedures). The Criteria and Procedures were established pursuant to a negotiation policy articulated by President George H.W. Bush, remained unchanged throughout the Clinton Administration, and thus remain unchanged in the current Bush Administration.

The Criteria and Procedures provide guidelines for the procedure for federal participation in negotiations to settle Indian water rights and criteria for determining federal negotiating positions. The process is designed to allow the United States to speak with one voice in tribal water right negotiations. To do so, federal teams are assigned to specific negotiations with members representing each major federal interest in the particular water basin. For example, the federal team participating in the Milk River negotiations included representatives of the Bureau of Indian Affairs, the Bureau of Reclamation, the Fish and Wildlife Service, the Department of the Interior's Field Solicitor's Office, and the Department of Justice. Despite this broad participation, the role of the team is to act as trustee for the tribes.

To do so, the team attempts to address agency differences. However, the team has no authority to choose between conflicting interests during negotiations if they are unable to find a solution that accommodates all federal interests. Only after an agreement is completed and presented at a higher level within the Department of the Interior can those decisions be made. This is often too late to provide meaningful federal participation in design of solutions. When tribal water rights are not involved, no official federal policy on participation exists, and the particular administration is often criticized for failure to present a united front. Two approaches to reform of the federal process should be considered. In Model 1, the federal government participates in negotiations at a level that allows rapid resolution of conflicting federal interests. In Model 2, each federal interest receives separate federal representation, with public airing of conflict.

Model 1 is based on the Truckee River negotiations. The federal team in the Truckee River negotiations represents all federal interests reflected in federal law. Rather than waiting until the completion of negotiations to take a position between conflicting views, the team chair has ready access to a decision maker at the level of an Assistant Secretary. This streamlined process allows real-time federal participation in negotiations. In doing so, it increases the likelihood that internal conflicts will be resolved and that the national interest will be reflected in a final agreement. Model 1 thus moves the level of federal participation in basin negotiations up a notch in the hierarchy of authority at the Department of the Interior from the level in the process under the Criteria and Procedures.

The Western Water Policy Review Advisory Commission recommended the appointment of a Cabinet level water official to resolve internal federal disputes. A position at that level is more likely to command the attention of Congress in shepherding a settlement bill through to final approval. However, it is difficult to argue that issues that are primarily western and within the domain of the Department of the Interior warrant this level of attention. Concern has been expressed that, rather than resolve issues, it could merely raise the conflict to another level by pitting cabinet level Secretaries against each other. In the context of Indian water settlements, the Chair of the Working Group on Indian Water Settlements with final authority to resolve internal disputes within the Department of the Interior is housed in the Secretary of the Interior's Indian Water

Rights Office. In the Truckee River negotiations (Model 1), decisions are made at the level of an Assistant Secretary.

In contrast, Model 2 would step down the ladder by eliminating the requirement that the federal participants speak with one voice. If one role of federal involvement in western water negotiations is to assure accountability to the public-at-large, efforts to speak with one voice may compromise that role. Although in a perfect world there should be no conflict in the implementation of federal mandates by various federal agencies, and no conflict with the federal role as trustee because Congress would have taken that into account, this ignores reality. These conflicts reflect differences in the interests of the public-at-large. Under Model 2, the public airing of these differences as they apply to a particular basin, rather than an internal effort to resolve them according to some monolithic national agenda, would not only ensure a broader public dialogue, but would increase representation of the divergent national interests in water use in a specific basin.

By airing the debate as it applies to a specific basin rather than attempting to fit diverse basins into the framework of a national policy, tailored solutions are not jeopardized. At the same time, a national role in the dialogue is not sacrificed in favour of a purely local solution. Achieving this federal process would require separate federal representation of separate federal interests in a negotiation. Thus, for example, representation of the interests of a federal water project must be separated from representation of a tribe. The difficulty with Model 2 is determining who makes the final decision between conflicting national interests once they are aired. In a water adjudication before a judge, the United States may aggressively assert the conflicting water claims of a tribe and a federal project or national wildlife refuge, leaving the final decision to a judge. In negotiation, that final decision must be made voluntarily. Thus, Model 2 must follow the full public airing of conflicting federal interests with a final decision.

There are additional weaknesses in the approach in Model 2. The public airing of conflicting national interests risks polarizing the local and national public, causing negotiations to grind to a halt. However, conflict on issues so important as to be reflected in conflict in national policy will ultimately surface. Better that they be addressed during the problem-solving phase of a negotiation than be raised in the context of an Environmental impact Statement once negotiations are complete and interests in the solutions are entrenched. Model 2 also increases the possibility that the benefits to all interests will be maximized through participation of a more comprehensive group of interests in the process.

Although Model 2 seems to solve the most problems with the federal process, ultimately a choice among the current approach, the two models, or some fourth approach should be made by the Department of the Interior in the context of a dialogue involving water interests and the general public. Of greatest importance is that the dialogue takes place and that the message be conveyed that the current process under the Criteria and Procedures is not working.

Political Leadership : The role of congressional leadership in the Truckee River negotiations and state legislative participation in the Milk River negotiations proved invaluable to finalization of the agreements. Leadership not only facilitated approval of agreements in Congress and the Montana Legislature respectively, but also provided an avenue for accountability to the larger public interest. Montana law mandates the participation of state legislators. The participation of Senator Harry Reid in the Truckee River negotiations was purely voluntary and, according to the parties, essential.

Assignment of staff for both Senators and Representatives from the affected state and staff for key congressional committees to participate in water negotiations from the outset would facilitate both approval and the public accountability of the process. Direct participation at the table may not be appropriate where negotiations are for the purpose of settling litigation and the federal interest is represented by the Department of Justice.

However, at a minimum, congressional staff should be informed of, and provide feedback on, solutions under consideration. Although congressional approval of a water settlement, once negotiated, does provide opportunity for review, it comes at a stage where substantive changes to solutions are unlikely to occur. The end result is either approval with relatively little change or failure to approve, neither reflecting real congressional input.

Similarly, participation of local legislators will facilitate consideration of the interests of citizens of the state and the state approval process. In addition, participation by tribal councils assures accountability to the tribal public. Although this early involvement of multiple interests results in upfront expenditures, the time and money spent on participation at a meaningful juncture would be well advised.

Equitable Distribution of Benefits

An equitable distribution of benefits from water use requires that water be allocated equitably among members of this generation, that the allocation reflect current values, and that the use, development, and health of a water basin be sustainable. Assuring an equitable distribution of benefits within this generation turns in part on efforts to develop tribal water. These efforts are discussed in the following section. Whether a negotiated solution promotes sustainable use of the water resource is tied directly to decisions on the future of major water projects in western basins. Most of those projects are federal.

The previous section recommended congressional effort to set priorities on this issue. This section recommends elements that should be considered in setting those priorities. Considerable environmental damage has resulted from the damming of rivers in the West. However, westerners are also fortunate that the previous generation invested in water infrastructure because now they have cheap hydropower, drinking water, and irrigation. The focus in setting priorities should be to remedy the harm, not eliminate the benefit. However, some federal projects serve production that cannot be considered necessary to the nation's food supply.

To prioritize the use of a federal project, Congress must first determine whether the benefits of cultivation of a particular water basin are adequate given the costs. Prior to expenditure of additional federal tax dollars, Congress should consider whether specific projects support a sustainable local economy or merely subsidize an artificial one. If the specific project subsidizes an artificial economy, additional congressional debate should take place on whether the product resulting from the subsidy is necessary to the health of the nation or whether it would be more appropriate to spend federal tax dollars assisting the local communities to convert to a more sustainable economy.

Second, if the benefits of a project justify expenditures, Congress should still consider in its authorization of funds whether improvements in efficiency, conservation, and alteration in regulated flows to mimic natural processes can reduce environmental costs. Although consideration of environmental impact will take place during environmental review of a proposed federal project, it is in the allocation of funds by Congress that changes to reduce environmental harm can be affected. An argument can be made that subsidy of western water use to facilitate these changes should cease, and those who have benefited should pay both the full cost of future water delivery and the cost to mitigate the environmental harm that fueled their benefit; however, this view oversimplifies the problem and its cause. The federal effort to dam the rivers and settle the river valleys of the West was undertaken in the national interest as it was identified in the early 1900s, and the current environmental harm is a direct result of that policy. By providing federal assistance to remedy that harm, Congress recognizes the reality that just as it required a national commitment to develop the rivers of the West, it will require a national commitment to address the consequences.

Paying for Solutions

Given a sufficiently deep pocket, it is likely that all of the West's water problems could be solved. Given reality, modern solutions to water problems must work within existing water infrastructure and attempt to design solutions that rely entirely on local water sources. However, if the United States is to meet the goal of providing an equitable distribution of water benefits, both within the current generation and between generations, then it must find the will to expend money in three areas: 1) development of tribal water resources, 2) repair to environmental harm caused by past development and use of water resources, and 3) renovation and efficiency improvements to existing water infrastructure. The previous section addresses funding for solutions to the latter two areas. This section addresses funding for development of tribal water resources.

Federal funding for development of tribal water resources requires two steps: congressional authorization of funding and appropriation of actual funds in the federal budgetary process. The Balanced Budget Act of 1985 placed caps on appropriations. Thus, even though funding may be authorized for development of Indian water rights, appropriation of funds for that development requires a reduction in another program covered by the budget of the Bureau of Indian Affairs. At the urging of the Ad Hoc Group on Indian Water Rights, Senator Pete Domenici (R-N.M.) introduced a bill on July 17, 2001 to allow a budgetary cap adjustment of up to $200 million per budget cycle to fund Indian water and land settlements.

The bill was referred to committee, and the Senate has taken no further action, the absence of a cap adjustment at the very least, or an even stronger commitment to establish a permanent fund for Indian water settlements, pits tribes and tribal programs against each other for limited federal funding. Until Congress takes this important step, the commitment of the United States to fulfil its role as trustee in developing Indian water and to assure the equitable distribution of benefits to water resources is in question.

Several professors and students from Lewis and Clark Law School and their associates recently participated in a relay run across America to raise money for research on muscular dystrophy. While traversing a lonely stretch of highway from west to east between the Sand Hills and St. Paul, Nebraska, at a pace between seven and eight miles per hour, they noticed that the change comes on slowly and is almost imperceptible at first. The air thickens, sweat lingers, there are fewer centre pivots visible in the fields, and the farms are closer together. Crossing this same region of the one hundredth meridian in the opposite direction, early settlers would have noticed it too, except heading West the air would become lighter and their lips and hands would have cracked.

John Wesley Powell understood what aridity would mean to development of the West. Over one hundred years since he recommended that governance of the West be established around its water basins, frustration over inability to administer and distribute water efficiently and fairly across jurisdictional boundaries has brought Westerners to the table to negotiate solutions that look a great deal like management based on basin boundaries. The rapid growth in the use of negotiation to solve chronic water management and use problems created by jurisdictional boundaries that do not follow watershed divides, water resources development that ignored environmental costs, and inefficient and inequitable distribution of the benefits of water use provides an opportunity for creative solutions.

The danger is that these locally driven processes will proceed without providing comprehensive solutions to these problems and without consideration of the long-term sustainability of the water resource and the economies dependent on its use. The time is ripe to examine these processes to determine the elements that can best lead to an efficient and fair process with a durable outcome. In doing so, negotiation can reduce costs, thus providing a more efficient process through joint collection and evaluation of data, and can allow development of more comprehensive and satisfactory solutions than are available through litigation.

The fairness of the negotiation process requires attention to inclusion of a widening circle of interests. People with direct interests in water must be represented at the table. People with an interest in the effect of water use on the basin community must also have an opportunity for comment at a time in the negotiation process when solutions are still in the design phase. The interests of the public-at-large must be represented by a national presence in negotiations, both in the form of federal representatives and congressional leadership.

Durability of the outcome requires that the process and solution be comprehensive by including all interests and all relevant issues. However, the complexity of a truly comprehensive solution warrants a phased approach in which manageable segments of the problem are solved at each phase. In addition, durability requires that the outcome result in an equitable distribution of water resource benefits both within the current generation and between generations. This requires both a shifting of development dollars to populations previously overlooked and attention to renovating existing water infrastructure and repairing environmental damage with an eye toward sustainable use. Finally, a durable solution must be one that anticipates the steps necessary for finalization by providing simple and flexible rules for administration and management that anticipate the likelihood of future changes in supply and demand, by providing funding, and by providing a mechanism for resolution of future disputes.

An efficient and fair process that results in a durable outcome is not impossible to achieve. Its precursors are taking place in water basins all over the West as illustrated by efforts in the Milk River Basin of Montana and the Truckee River Basin of California and Nevada. In order to assure an appropriate solution, as a nation we must commit to ensuring the sustainable use and flow of the rivers of the West.

2

Water Reform in Southern Africa

At the September 2002 World Conference on Sustainable Development, held in Johannesburg, South Africa, water was at the centre of contestation and debate. Water was variously characterized as a scarce resource, an economic good, a human right, a matter of national and international security, and an environmental right. In this chapter, we examine the process of water reform underway in Zimbabwe.

The complex interplay of environmental, economic, social, and rights-based discourses and practices related to this essential natural resource provides a terrain from which to consider current debates surrounding how the "political" is conceptualized in political ecology. First, our study draws attention to the multiple ways water is "valued" in international, national, and local discourses and how these different dialogues are used by actors to position themselves and their interests in Zimbabwe's water reform process.

Second, the case study raises questions concerning the liberatory nature of Zimbabwe's supposed populist political agenda in land and water reform. While the reforms ostensibly aim to increase racial equity by providing black communal area farmers with access to water, the emerging realities on the ground are quite different. As with Zimbabwe's "fast-track" land reform program, it appears that the major beneficiaries will be middle-or upper-class black entrepreneurs and political supporters of the Zimbabwean African National Union-Patriotic Front (ZANU-PF) rather than black land-and water-short communal area farmers.

The notion of "people" thus is a highly contested category in present-day Zimbabwe. We begin by outlining our conceptual framework and describing the methodology used in the study. While set in a political ecology framework, we draw on Arce and Long's (2000) concept of counterdevelopment to illustrate how local understandings are used to comprehend and reshape global discourses and how, in turn and less frequently, the local influences global discourses. We then present the case study of water reform in Zimbabwe as a means of demonstrating the grounded, decentered, and engaged approach of political ecology.

Focusing only on one pervasive discourse, such as neoliberal economic policy or the growing scarcity of water, and studying its effects on people and the environment, misses much of the complexity embodied in the reform. We identify different and often competing actors and describe their conceptualizations of the value of water. Our emphasis thus draws attention to the role of multiple actors, history, ambiguities, and contestations.

Political Ecology: Research Issues and Methods

This chapter was an interdisciplinary, collaborative, and comparative project involving faculty and graduate students at the Centre for Applied Social Science at the University of Zimbabwe; researchers and graduate students at Chancellor College, University of Malawi; a colleague at the Kennedy School of Government, Harvard, and ourselves at Michigan State University. Given the interdisciplinary nature of the research, political ecology proved to be a particularly useful approach, as it provided a shared analytical procedure that encompassed the issues and methods familiar to the anthropologists, resource economists, and historian engaged in the study. Since researchers from different disciplines were involved, key concepts were presented directly, avoiding as much disciplinary-based terminology as possible. The concepts were derived from the works of Blaikie (1996, 1999), Blaikie and Brookfield (1987), Stott and Sullivan (2000), and other political ecologists. Researchers agreed on the following conceptual framework:

1. The central concern of political ecology is understanding the relationship between social and environmental change. Political ecology draws on insights from a variety of environmentally related disciplines in the social and environmental sciences.
2. The outcomes of environmental change are often felt unevenly by different social groups. Explaining why and how this unevenness is generated links political ecology to political economy and makes conflict and contestation over resources

central to most analyses. Contestation involves studies of struggles at the ideological and discursive levels including, but not restricted to, definitions of science, knowledge, environment, sustainability, and biodiversity.

3. Power is a central focus of the political in political ecology. Increasingly, a concern with power relations extends beyond the local level and decenters and problematizes unidimensional treatments of the state, donor groups, nongovernmental organizations, and their related discourses.
4. In contrast to earlier approaches, which assumed that ecological systems tended toward equilibrium, political ecology recognizes that resource utilization patterns may be ecologically degrading while being socially profitable or functional, at least in the short term, for some actors.
5. Ecosystems and social systems are regarded as mutually constituted. In theory, the goal is to develop modes of analysis that encompass and relate social and ecological variables, although as noted there is an increasing tendency to focus analysis on social and political factors and to pay less attention to the ecological variables.
6. Political ecology combines and relates different levels of analysis. Conceptualization of these levels requires new styles of analyses since much of the local is permeated by and can mirror and refract the global. Similar complexities exist in linking microhabitats and bioregions to global environmental change and to social factors operating on different scales.
7. Political ecologists usually study the complex interactions between a changing environment and changing society within the context of local histories and ecologies. While Peet and Watts (1996) lament the absence of broader theories, political ecology has tended to yield historically and culturally contextualized conclusions.
8. Most political ecologists share a concern with policy formation, social justice, and a linking of research to action.

This chapter reports on the research findings from Zimbabwe. In line with the focus of this special issue, we concentrate on the political and power dimensions of water reform and the research methods used in the study. The overall research goals were to examine if and how Zimbabwe's new water reform broadened disadvantaged groups' and women's access to water and increased their voice in the new institutions and laws established to manage water resources. More specifically, we asked, would the new water laws and policy help raise the poor's standard of living? Would black and white farmers be able to work together as members of the new water institutions to achieve new goals? Would women and their concerns be incorporated into the new institutions and practices? Would new patterns of democratic decision making be created outside the rigid party structure of the ZANU-PF, which until 2000 had been the dominant political party and virtually the only party in rural areas? In short, we placed the political at the centre of our study with a deep concern for social justice in policy formation and implementation.

We adopted a proactive stance in the research process and provided a continuous flow of recommendations for increasing marginalized groups' participation in the newly instituted Catchment and Subcatchment Councils and other management bodies. In the long run, the research goal is to ascertain what environmental effects the reform has produced, as reflected in changes in water use, distribution, quality and quantity issues, as well as the social, economic, and political ramifications of these changes.

Although most countries in Southern Africa are undergoing similar transformations in their water sectors, Zimbabwe was selected for study because the reform was further along in implementation than reforms in many other countries in the region. We designed a multipronged and multisited research strategy that included the following elements:

1. Review and analysis of international documents and policy statements on water resources, including World Bank (1993, 2001, 2002), international donor policies and projects, Food and Agriculture Organization of the United Nations (FAO 1995, 2000), World Water Council (2000), World Water Forum (2000), and a review and analysis of Government of Zimbabwe policy and project documents and new laws.
2. Attendance at major planning and training workshops held over four years on Zimbabwe's new water policy, acts, statutory instruments, and principles of water allocation. Regular participation over this period in the new stakeholder water management institutions, catchment and subcatchment councils, in the Mupfure and the Mazowe Catchments and later in the Sanyati and the Manyame Catchments.

3. Closed and open-ended interviews with donors and policy makers internationally and nationally. Intensive interviews with stakeholders on catchment and subcatchment councils in our research areas.
4. Focus groups at the village level on water sources and use and knowledge of new policies and laws in the three catchments.
5. Surveys of water use and water investments in selected villages and commercial farms in the research areas, which generated baseline information on water use and investment and examined if and how the water reform was addressing critical rural livelihood issues.

In part, we relied on Arce and Long's (2000:19) concept of counterdevelopment, which they describe as involving a "balancing act between introduced bureaucratic preferences and procedures and local practices." This approach recognizes that development and modernity assume varied faces and forms as they are translated from the centre outward.

The Political Ecology of Water Reform in Zimbabwe

Our case study of water reform in Zimbabwe illustrates a problematic aspect of Vayda and Walters' (1999) call to identify environmental change as the starting point of analysis. The study calls attention to the importance of a situated, historically aware analysis that considers the political, discursive, and ecological to be mutually constituted.

We began the research by focusing on the social, political, and policy dimensions of Zimbabwe's water reform, with the goal of later examining the reform's social and environmental consequences. Although the political was foregrounded, this was not because we downplay the importance of environmental change. Instead, it reflects that the reform is in its early stages and to date has had little environmental impact. Waiting to begin study until the environmental consequences are evident at the local level casts political ecologists in the role of carrying out analyses of impacts and negates their part in guiding policy and practice. It is in the nexus between the social and the ecological-in this case with priority given to the political-that we can begin to understand why Zimbabwe's water reform is constituted the way it is and what its probable social and environmental consequences may be. This perspective permits examination of the interplay of the varied conceptualizations of the value of water and the different understandings of water reform at the international, national, and local levels.

Multiple Discourses on Water

We first identified competing ideological frameworks of valuing water in Zimbabwe and Southern Africa. While it is tempting to present them as independent of one another and subsumed by an ascendant global capitalism espoused by institutions like the World Bank, what existed was more complex. Various actors in Zimbabwe's water reform made use of these frameworks in different ways in attempting to shape the new policy document, laws, and practices to serve their interests.

At the international level, the Dublin Principles are widely recognized axioms for water reform, and the thinking behind them has been incorporated into policy documents authored by the World Bank and other donor organizations (World Bank 1993, 2002; FAO 1995, 2000). The four principles are: 1) fresh water is a finite and vulnerable resource, essential to sustain life, development, and the environment; 2) water development and management should be based on a participatory approach, involving users, planners, and policy makers at all levels; 3) women play a central part in the provision, management, and safeguarding of water; and 4) water has an economic value in all its competing uses and should be recognized as an economic good.

These four widely espoused principles reflect more than a single economic discourse or recognized rationale for water reform. Like all international documents, the Dublin Principles were shaped by the varied interests of the policy drafters. For example, the water policy advisor for NORAD (the Norwegian development agency) was instructed to ensure that water was treated as an economic good as well as to recognize women's roles as water managers (personal communication with Dr. Desmond McNeill, NORAD water policy advisor, November 2001).

International documents reflect other competing discourses on the value of water which, if not yet dominant, may become so. Peter Gleick (1999) adopts a human rights approach. He views water as different from other commodities and argues that, with growing global water scarcity, a right to water should be recognized to protect the poor and vulnerable from having an essential ingredient of life priced beyond their means. Gleick's rationale is similar to that used by those who promote a human right to food (Eide 1995). In an earlier paper, we also proposed a human-rights-based approach to water reform (Ferguson and Derman 1999). Using the framework of the Convention on the Elimination of All Forms of Discrimination against Women, Helium (2001) suggests ways for water planners to legally recognize women's rights to water and to include them in planning and policy making. With the growing importance of a human rights approach to development, a rights-based approach to water may ultimately rival a commodity-focused one.

Another important orientation to water draws on the metaphors of scarcity and security. Security issues can be soft (environmentally focused) or hard, with an emphasis on the role of water scarcity in war and conflict. This approach is reflected in an FAO (2000) document that develops the concept of freshwater security and in de Villiers' (1999) focus on the prevention of war. Finally, water use and conservation principles are found in international policy-setting environmental documents linking water to the global agenda for sustainable development. The chapter on water resources in Agenda 21 is an example, as are the documents and policy statements emerging from the most recent environmental summit in Johannesburg, South Africa.

Those who provide water as well as those who use it are engaged in struggles to redefine rights of access to the resource. This is reflected in shifting conceptualizations underlying the Rural Water and Sanitation Programs in Africa. Cleaver (1998a, 1998b), Cleaver and Elson (1995), and Ferguson (1998) have documented this change in approach and have explored its ramifications for women's control over water resources. Water is now frequently conceptualized as either "productive" or "primary," with greatest importance in most policy documents given to its productive uses. A similar change in thinking about potable water has occurred in the public health domain and in water management circles (Nicol 2000).

South Africa is pioneering an effort to shift the focus on water resources from a supply-to a demand-side approach. Water demand management is also widely promoted in Southern Africa by the World Bank and the International Union for Conservation of Nature and is often promulgated by water engineers with seemingly little concern for its social consequences. The underlying principle is based on the scarcity model. While the supply-side approach considered water as a public good, demand management requires quantifying the amount of water available and then managing it within these limits using pricing options and other measures.

At the same time, however, others within Southern Africa are using other conceptualizations of the value of water based on the belief that water management and rights should be used as a means of environmental and social transformation and justice. South Africa's new Water Act (1999) reflects this approach as well. The point is that these various discourses all coexist in differing combinations from country to country, institution to institution, and among different actors within the same country or institution. Many are informed by neoliberal economic thinking, but others respond to counterdiscourses and histories. States in Southern Africa are constructing their national water laws and water management administration, often drawing on multiple (and sometimes contradictory) frameworks. Actors within these countries use various frameworks to further specific political and personal agendas.

Zimbabwe's Water Reform in Global and Local Context

How are these various conceptualizations of water-as a scarce commodity, as a human right, and as a tool for social transformation and greater participation and equity-being used in Zimbabwe? How do different actors use these international frameworks and more locally derived ones to attempt to shape and benefit from the reform? As part of the process of selective incorporation and use of discourses, we consider how certain actors seek to avoid the application of particular international principles or to refashion them to suit other ends.

Zimbabwe's newly enacted water reform policy and laws contain and sustain contradictory principles and agendas. For example, the concept of primary water was incorporated into these documents, thereby acknowledging that people have a right to use water for drinking, cooking, washing, watering livestock, making bricks for houses, and other noncommercial purposes. However, these same instruments embrace the key neoliberal concepts of water demand management and cost recovery espoused in the Dublin Principles, the World Bank, and other large international lending and donor institutions. Left out of Zimbabwe's new policy and laws, but debated at the local levels, are other perspectives on water. For example, one widespread discourse throughout Southern Africa is the close relationship thought to exist between ancestral spirits, the state of society, and rainfall. Conceptualizations of water embodied in the new laws and policy do not include consideration of ancestors' agency in producing rain based on their evaluation of the moral state of civil society! Nonetheless, these struggles over meaning can be just as important as the struggles over the resources themselves (Derman and Helium 2002; Moore 1996:128).

These competing discourses include assumptions about the nature of water, the hydrological cycle, a changing environment and economy, the relationship between water quality, people, and disease. Embedded in the different perspectives are assumptions about the monetary, political, social, and ecological value of water. To account for this complex set of competing understandings within a political ecology framework that privileges environmental impacts over a more inclusive and iterative approach to human-environmental interactions, or to suggest that the discourse and practice related to capitalism and neoliberal economic values are the prevailing ones, inadequately represents the current state of affairs and obscures people's agency.

Finally, the current political situation in Zimbabwe calls attention to the importance of the wider political, economic, and social context in understanding how the water reform process evolves. The implementation of the new water acts and policy is caught up in, and cannot be considered separately from, the current crises, involving the new "fast track" land reform. Since the failed constitutional referendum in 2000, most white-owned commercial farms in Zimbabwe have been resettled under the fast-track land reform program. Fast track is best characterized as land confiscation or appropriation by the national government for political ends. It represented one of ZANU-PF's and President Mugabe's strategies to win the March 2002 presidential elections. Through the elections, government vehicles were ferrying ZANU-PF supporters, war veterans, and others to occupy designated white commercial farms. Every law concerning land was violated, and laws were changed retroactively to make the process legal and to block the commercial farmers from obtaining legal remedies. The implications of this situation for the water reform process.

The New Water Laws and Administrative Structures

Zimbabwe's reform incorporates two key components of international neoliberal capitalist thinking about water: scarcity and participation. Notions of existing or impending water scarcity underlie the global rationale for reform, while decentralized, stakeholder management is regarded as central to improved resource use and conservation. These two concepts articulate with Zimbabwe's colonial history and racialized present, as reflected in government land and water policies. The motivations for water reform in Zimbabwe reflect an interplay among international forces, including the World Bank, United Nations organizations (especially the FAO), international environmental organizations, and different elements within the government. Central actors within national government are the Department of Water Development; the Ministry of Agriculture, Finance and Planning; and the President's Office.

New water policies and laws represent high-level Zimbabwean bureaucrats' efforts to use international principles in forging a national strategy to attract funding from the World Bank, FAO, and other national donors in line with their political, social, and personal agendas. Although local groups were not involved in the initial phases of the reform design process, key stakeholder groups were identified as a result of this process. These were large-scale commercial farmers represented by the Commercial Farmers Union, mining companies, cities and the Agricultural Rural and Development Authority, a parastatal which owns multiple farms throughout Zimbabwe, and government itself.

Following extensive consultation with the World Bank, internal meetings of stakeholders in Zimbabwe, and the report of Zimbabwe's Land Tenure Commission, which indicated that water distribution and use had the same inequitable allocation patterns as land did, the Government of Zimbabwe decided upon a restructuring of the water sector. The changes are embodied in two acts: the Water Act of 1998 and the National Water Authority Act of 1998.

These laws rest upon and draw from several not necessarily compatible ideas. Historically, one primary mission of the Departments of Water Development and Hydrology was to provide water to users. In the past, the Department of Water Development constructed reservoirs throughout the country in a manner that overwhelmingly favoured the white residents of the (at that time) colony of Southern Rhodesia who occupied the most productive lands. Because of periodic droughts, Southern Rhodesia was viewed as water scarce, requiring an extensive system of dams and reservoirs to ensure water supply for cities, mines, and farms.

The 1998 water legislation transferred national planning functions to a new parastatal agency, the Zimbabwe National Water Authority (ZINWA), which has become the owner of public dams that store more than 5,000 megalitres of water. ZINWA is to be funded primarily through the sale of water behind government dams, the provision of water to cities, and the levying of water to large-scale users. It will draft master plans for the development of Zimbabwe's waters and for the protection of its environment, and it will work with each of the seven catchment councils established as a result of the reform to develop catchment outline plans which include basic demographic, hydrological, and economic information to permit efficient water management, utilization, and development.

Localizing Water Reform in Zimbabwe

The research focused on how three national concepts have been refashioned as a result of the reform: the notion of scarcity, the concept of stakeholder participation, and the user pays principle.

Scarcity

In most global environmental reports, the growing scarcity of fresh water is regarded as a critical problem. The meaning and degree of scarcity in Zimbabwe, however, varies significantly by actor and by region of the country. Except in drought years, Zimbabwe has not suffered from an inadequate amount of water for its needs, although there is growing evidence it will face water shortages in the future.

Scarcity is as much a social as it is an ecological concept and can be "objectively verified" in social as well as ecological terms. For planners in Zimbabwe, the most important type of scarcity to address was not ecological, but social-increased access to water for black farmers. The historical factors underlying the land and water reform in Zimbabwe are the racial imbalances between whites and blacks. Irrigated agriculture consumes 70 percent of the surface waters that are used yearly. Until recently, the country's agriculture and much of its economy centred upon a large-scale commercial agricultural sector that was overwhelmingly composed of white farmers. These farmers held almost 50 percent of the land, although with recent land invasions and the new land reform process, their holdings have now been reduced to around 10 percent. Whether the current reforms will actually broaden black communal area farmers' access to land and water resources remains an open question.

Local Assumptions of Water Scarcity and Distribution

In what may well be a mistaken assumption, the Zimbabwean government believed that systematic discrimination was taking place against black farmers as a result of the priority date water allocation system (PDS) established in the colonial period. It is likely that this system originally was chosen for its simplicity, its compatibility with water measurement techniques in the early 20th century, and as the political means whereby early farmers could protect their investments and farms. The PDS also established the legal framework to block African farmers from competing with Europeans. Over time, however, it caused far more conflict among white farmers than between them and black farmers.

PDS refers to the practice whereby those who obtained legal water rights earliest had the first call upon water in a river and held those rights in perpetuity. This led to continuous disputes between upstream and downstream farmers since the logic of the system was that the highest reservoirs were filled first in case the rains failed. A central feature of the new legislation, in contrast, is the institution of a permit system granting water rights for 20 years. On the surface, it appeared that rights in perpetuity severely limited new entrants into the water rights system and thus maintained racial inequities. This impression was heightened by white farmers' vehement opposition to abolishing the PDS.

However, the PDS is not the only way water was inequitably distributed. Government has stored massive amounts of water behind dams it constructed before and after independence. A significant factor limiting black farmers' access to water is their lack of capital to pay for irrigation infrastructure, not the PDS system per se.

The new fast-track land reform program, short on donor support and funding, does not at this time provide new settlers with credit, training, and other resources necessary to resume irrigated production on the commercial farms and, hence, fails to address some of the basic roots of racial inequality. The notions of scarcity and the call for greater equity in access to water thus are more than environmental facts; they are powerful political motifs in Zimbabwe.

Stakeholder Participation

The assumption that users will manage water more effectively if they are engaged in the institutions of water management is a central feature of the Dublin Principles and of Zimbabwe's water reform. How the Dublin Principles are applied varies from country to country-no formulas exist that identify who should be included in the new management institutions. In July 1997, two pilot catchment projects were launched in Zimbabwe, the Mupfure and the Mazowe, to study different models of stakeholder participation to determine which one should serve as the model for the country as a whole.

The Mupfure Catchment experiment was donor funded and government supported, while the Mazowe Catchment was privately initiated and organized. The Mazowe Project was strongly pro-business and opposed heavy reliance upon government on two grounds-the first related to government's lack of accountability and the second stemming from their belief that ultimately people had to rely on their own skills, abilities, and resources if the reform were to succeed. Mazowe Project planners tended to minimize the resource differentials among different stakeholder groups in their catchment, including those between white commercial farmers and communal area farmers. The Mupfure Project, in contrast, was far more government oriented and prone to making claims about how the water reform would result in greater prosperity and access to water. They relied heavily on donor funds to underwrite activities and to draw support, thus reinforcing an attitude of dependence on government. The Mupfure Project sought to engage communal-area farmers through meetings and locally based development projects, although donor support for these local level initiatives was short-lived.

In addition to these basic differences in philosophy, other key differences between the Mazowe and Mupfure rested in the definition of stakeholders and the proposed systems of representation. The Mazowe Catchment developed a bottom-up management approach. It instituted water-user boards (WUBs) as the third and lowest administrative tier and the location from which all decision making would originate. WUBs were composed of 15 members, 8 elected by users and 7 nominated to represent the most important economic sectors. The Mupfure, in contrast, did not have this third-level administrative tier and recognized instead only the catchment and subcatchment councils.

In the end, the government did not await the outcome of this experiment in representation and instead opted to impose its own structure and system of representation upon all catchments. The motivation was Zimbabwe's deepening political and economic crisis, necessitating a rapid shift of the financial costs of water management from central government to the new parastatal, ZINWA, via reliance on the "user pays" principle. Thus, ironically, centralized government chose the decentralized model to be followed in the catchments established as a result of the new legislation.

Localized Definitions of Stakeholder Participation

Perhaps even more significant than the transfer of authority to ZINWA is the creation of catchment councils and subcatchment councils as major management entities composed of representative stakeholder groups. Zimbabwe has been divided into a total of seven catchments, ecologically derived administrative divisions that do not correspond to other existing political boundaries and which create jurisdictional conflicts. Provincial governors and administrators, for example, do not feel ownership for catchments, as many transcend provincial boundaries.

According to the Water (Subcatchment Councils) Regulations of 2000, the following stakeholder groups are to elect representatives of the subcatchment councils: rural district councils, communal area farmers, resettlement farmers, small-scale commercial farmers, large-scale commercial farmers, indigenous commercial farmers, urban authorities, large-scale miners, small-scale miners, industry, and any other stakeholder group the catchment or subcatchment councils may identify. The number of members on subcatchment councils is not fixed by legislation, nor are the procedures for how stakeholders are to be selected from within their constituent groups. To date, no subcatchment council we are familiar with is actually composed of representatives from all the above-listed stakeholder groups.

Catchment councils are made up of the chairpersons of subcatchment councils and other members. In practice, and in contrast to what is called for in the statutory regulations, each subcatchment council sends two officers to the catchment council. There are no provisions for representation by stakeholder groups on the catchment councils themselves. The assumption is that stakeholders are represented at the subcatchment council level.

Catchment council responsibilities include: collaborating with ZINWA in preparing and updating catchment plans; deciding on and enforcing all water allocation and reallocation; developing and supervising programs for catchment protection; issuing and overseeing permits for water use; establishing and maintaining a database and information system (together with ZINWA); and overseeing operations and functions of subcatchment councils. Subcatchment councils are to monitor the allocation of water permits, water flows, and use. They also are to assist in pollution control, catchment protection and data gathering, and collect fees that will be used for the performance of their duties.

Race is an important element in participation. A central aim of the water reform was to broaden stakeholder representation beyond white commercial farmers by including black communal area farmers, members of rural district councils, and indigenous commercial farmers on the catchment and subcatchment councils. While black communal area farmers are now represented on councils, significant disparities remain between the number of stakeholders and the number of stakeholder representatives. The largest number of water users by far is communal-area farmers. Yet they have the same number of representatives on councils as, for example, large-scale indigenous farmers, whose numbers are very small in comparison. If there is to be a significant change in access to water, then representation patterns will also have to be altered.

While greater racial equity has been a centerpiece of the reform, gender equity, featured in the Dublin Principles and in many other international documents, and strongly promoted by international donors, has received little local support. This is despite women's heavy involvement in agricultural production and household water provisioning. Neither in the laws nor in the statutory instruments implementing the water reform has any mention been made of how to incorporate women more fully into water management and development. Observations at catchment council meetings indicate that the few women present are usually either technocrats or secretaries. Few women have been elected to councils.

Finally, our surveys indicate that key stakeholders identified at the village level are not represented on councils. Seventy-five percent of rural Zimbabweans interviewed believe that customary authorities-chiefs and spirit mediums-should regulate water. Yet, they have no formal representation on stakeholder bodies. In general, no clear-cut divisions were observed between black and white participants on catchment councils.

The commercial farmers and the government technocrats often speak a common technical language. They are familiar with concepts like water rights, permits, flows, and storage rights, but the white farmers are highly suspicious of government and do not trust that monies collected will be used for stated purposes. They share a sense that the national government has been highly corrupt. While discussing such issues, participants usually take care to assure that the government officials present are not part of the problem. White commercial farmers and members representing communal areas share an intense skepticism of government capacity, and they often agree that concerns of communal area residents are ignored by government.

User Pays

The Zimbabwean application of the principle of water as an economic good is "user pays." This principle is consistent with Zimbabwean notions of commercial water and is central to the functioning of the new administrative structures. User pays has become a rallying point various actors use to garner support for their versions of water reform. User pays derives its authority from the Dublin Principles, although the drafters might not recognize it in such bald, free-marketeer form. While capturing the reductionist philosophy of those seeking to make water an economic good and reducible to market forces, the application of this principle in Zimbabwe indicates the need to move beyond international policies and power relations to discover how what appears to be an axiom of all reform is shaped by various actors. Thus, research reveals the more complicated national and local agendas behind the international economistic perspectives advocated by the World Bank and other major lenders and seemingly adopted whole face by the government of Zimbabwe. Similar to other policies, this one is negotiated, reworked, and presented differently to various national constituencies.

The vast majority of Zimbabweans does not have water rights in the legal sense of the term and are not likely to gain them as a result of the new laws. What they have are primary use rights that enable them to use water without payment for drinking, washing, watering livestock, and tending small gardens. The principle of user pays applies to only the small number of consumers (mostly large-scale irrigated farms and mining operations) who use water for "commercial" purposes. These commercial users, while small in number, nonetheless consume most of Zimbabwe's water. Commercial is not defined, but commercial uses are specified as agricultural, industrial, urban, electrical, and includes a category of miscellaneous for all those uses not defined. Although the number of commercial users is not great in comparison to the total population, neither ZINWA nor catchment councils have the capacity to monitor and enforce the user pays principle even among this small group.

In emphasizing the principle of user pays, water reform goes against how Zimbabweans view and value water. In village-level surveys in three catchments, we found that over 90 percent of those interviewed did not think they should pay for water used for commercial purposes. When payment is involved, the consensus was that it should be for irrigation, canals, pumps, and other infrastructure, not for water per se. In short, user pays is not a principle accepted by most rural dwellers. The new laws do not affect urban users. While the majority of urban households use water for "primary" purposes, they are increasingly being charged for water consumption. Little attention has been paid to the urban and periurban dimensions of the reform, although many Zimbabweans now live in these areas.

ZINWA and the catchment councils are further handicapped because there is no mechanism to determine what constitutes either a sound economic price for water or a socially acceptable one. Despite this situation, these new administrative bodies are to be funded through the sale of water and levies upon water permit holders. For the moment, ZINWA has set new prices for water in an effort to generate the needed revenue. Great uncertainty exists, however, concerning commercial farmers' abilities or willingness to pay, as many large-scale commercial farmers have either been forced off the land or are refusing to pay for water because of the threats against their farms. As we write, discussions with those sent out to collect payments and levies from the new black settlers-those now living on what had been commercial farms-indicate either that they are not farming or that they are not willing to pay for water. It is not clear how ZINWA will survive given these conditions. Here clearly, the linkage between the land and water reforms is evident, even though there has been little effort to relate or coordinate these reforms of Zimbabwe's most fundamental natural resources.

Another contentious feature hampering the water reform is that the former staff of the Department of Water-engineers, hydrologists, accountants, planners, secretaries, drivers-all remain employed, now by ZINWA instead of the Department of Water. They have been transferred to and now serve as the catchment manager's staff. One catchment alone has 62 employees. The funding implications of this transfer of personnel (amounting essentially to the creation of seven mini water departments), along with the issue of the emerging working relationships between "experts" and stakeholders, constitute part of our ongoing research.

The placement of large numbers of government employees and technocrats in stakeholder organizations is indicative of government's suspicion of the newly created stakeholder organizations. Government's role has become a central point of contestation in catchment and subcatchment councils. The following excerpt from a Mazowe Catchment Council meeting, April 24, 2001, illustrates this point. Mr. Chinamhora, a distinguished older man representing a communal area, who has been active on the Mazowe's Catchment Council since its inception, asked a question which had been answered several times before, reflecting stakeholder dissatisfaction with the current arrangements: "Are officials in the catchment manager's office ZINWA employees or catchment council employees?" The catchment manager responds, once again, that "They are ZINWA employees paid by ZINWA."

When Mr. Gotora, the Mazowe Catchment Council chairperson, pressed the officials to reveal who gave them contracts, it emerged that they still did not have signed contracts. Gotora assured them that the catchment council (CC) would support their efforts to obtain contracts and maintain their jobs. This discussion resumed a little later with a slightly different emphasis. Chairperson Gotora asked the catchment manager, "Where do permit application fees go?" The catchment manager responded, "They go to ZINWA because they are the final permit issuing authority." Gotora then stated, "If the application fees go to ZINWA, why should the CC process permits? Why should ZINWA get paid for work done by the CC?" To this the catchment manager gave no response. These are matters that are not addressed in the Water Act but which rather constitute the "details" of implementation. They require that stakeholder groups become self-financing at the same time they are generating information and revenues for ZINWA.

A central concern of commercial farmers and others is: "What will we get in services for our monies?" This is a potentially divisive issue between commercial and communal areas in catchments because those not holding water permits may not be entitled to any services. Some commercial farmers assumed that since they were paying for permits, they should be the ones reaping the benefits. In the end, however, some commercial farmers and others argued that unless water reform had a developmental dimension benefiting communal area farmers as well, there was no point in carrying it out.

Nonetheless, a fundamental divide remains in how water is valued. Catchment and subcatchment council members have learned that without the user pays principle, they will not have the resources to function. On the other hand, communal area farmers who have not had to pay for water in the past continue to resist notions of paying for it. Communal area farmers surveyed were divided evenly over the issue of whether water should even be regulated. They were almost unanimously opposed to paying for water for primary purposes (drinking, watering livestock, and brick making). While the new water management institutions may be comfortable valuing water monetarily, most small-scale farmers continue to resist this approach.

The Current Context

These issues must be placed within the wider context of the implosion of the government budget and the economy as a whole since 2000. Zimbabwe's economy is in free fall. It is characterized by soaring inflation, lack of external investment funds, exceptionally high internal rates of interest driven up by the government's borrowing, a foreign exchange shortage, an unbudgeted war in the Democratic Republic of the Congo, and unappropriated expenses to raise salary levels for potential supporters of the president.

The crisis has been intensified by the withdrawal of major lending and donor organizations, including most notably the World Bank, the European Union, and the government of Denmark. Those donors that remain supportive are shifting their emphasis to supplying food for an increasingly impoverished and hungry nation. With the exception of the Dutch government, which continues to support two catchment councils (the Sanyati and the Mzingwane) and the creation of an Association of Catchment Councils, donor support for water reform has all but disappeared.

The current drought means that there are growing shortages of water while the capacity to manage water has been diminished due to the financial crisis. Thus water shortages now pose a national crisis. The proposed solutions are to continue to shift water management away from government to catchment councils, subcatchment councils, and a self-funded ZINWA, despite lack of potential revenues at the local level. The existing climate of violence and land occupations renders the prospect of successful water reform highly unlikely. To summarize, using illustrations from Zimbabwe, we have drawn attention to the multiple ways water is "valued" in international and national contexts and to how these different dialogues are used by actors to position themselves and their interests.

We have also demonstrated that the "emancipatory" rhetoric and potentials of the reform are not as straightforward as one might think. While the reforms ostensibly aim to increase racial equity by providing black communal area farmers with greater access to water resources, the emerging realities on the ground are quite different. As with the fast-track land reform program, it appears that the major beneficiaries of water reform are likely to be middle-or upper-class black entrepreneurs and political supporters of ZANU-PF rather than black land-and water-short communal area farmers. Women and their diverse interests also have been left out of the reform, despite the international rhetoric.

The notion of "people" is a highly problematic category in Zimbabwe, where the government and the party claim to speak for the people while excluding all "whites," all members of the opposition political party, and almost anyone who disagreed with ZANU-PF's program for the reelection of Robert Mugabe. Of the multiple principles associated with water reform both in Zimbabwe and internationally, three constitute the keystones of Zimbabwe's reform: notions of scarcity, stakeholder participation, and the user pays principle. Surveys, carried out in several areas of Zimbabwe since the initiation of pilot catchment reform schemes in the mid-1990s, indicate that user pays is a highly contested way of valuing water. With the exception of the infrastructure in irrigation schemes, most Zimbabweans do not think they should pay for water. Thus, the government and the newly established catchment councils have positioned themselves against the majority of the population's understanding of water.

The research also raises questions about the assumed privileged role of the poor in environmental conservation, development, and water management. Management of Zimbabwe's water resources in the seven catchments necessarily entails reliance on the multiple perspectives, including those of large-scale commercial farmers, hydrologists, ecologists, and numerous others, in addition to poor communal area farmers. While highly participatory and democratic frameworks may suffice in small watersheds, the large size of Zimbabwe's catchments and their international dimensions pose more complicated issues of representation, voice, and expertise.

Our experience working with development for many years in Africa in diverse contexts leads us to share Arce and Long's perspective on how dominant discourses and practices are reshaped in different national and local settings. Political ecologists using this framework would inquire into how local understandings, or in Martinez-Alier's (2002) words, languages of valuation, are used to comprehend, reshape, or resist these global processes.

Water reform has taken back stage (for the moment) to national politics in Zimbabwe, but, given the reappearance of serious drought, contestations over water will again intensify among users in different economic sectors and with differential access to water. The old systems for managing water are no longer functioning, due to the land invasions, while the new systems are not in place. This means that the years of careful planning and implementation of water reform are now in jeopardy. We wish the power balances in those contestations had been altered so that the vast majority of water users would have greater say and greater access to the new water-related institutions. Under the current circumstances, we remain skeptical that the government has any intention of doing so.

The current political and economic crisis taken together with the complex world of water reform, including multilateral banks and institutions, national water departments, engineers, large-scale farmers, women irrigators, and other stakeholders, makes finding social and environmental policies that work an even more distant goal. Large-scale experimentation in the water sector is well underway throughout Southern Africa, with relatively little vision of the possible social or environmental outcomes and with what sometimes appears to be fatuous optimism. Political ecologists, with their focus on the dynamics of human environmental interactions, their concern with power relations, and the linking of theory and practice, are well situated to study this process and to provide recommendations for the way ahead.

3

Wellsprings: Water Regeneration

Based on long-term ethnographic research, this chapter presents a comparative view of two groups in Queensland: a local catchment management group in Brisbane, and the participants in a bi-annual water festival in Maroochydore. Both are composed of social networks whose purpose is to engage with local water issues and participate in the management and protection of local waterways. In doing so, both groups articulate notions of connection and belonging, and collective responsibility for environmental care. However, their visions of what constitutes care, and their forms of knowledge and discourse differ considerably.

Through the ethnography, this chapter attempts to show how their particular approaches serve to reveal some of the underlying differences in social and environmental knowledges and values which underpin wider debates on water issues in Australia. These differences are both intellectual and political. At a conceptual level, they make use of somewhat differing models of human environmental relationships, on the one hand subscribing to a largely technical model in which humankind and 'nature' are considered dualistically and, on the other, making use of a more holistic model, in which humankind is perceived as being within and integral to 'nature'.

As Descola and Palsson have made plain (1996), the former vision is one that dominates the environmental relationships of Western societies, and reflects the abstraction and conceptual compartmentalisation of large industrialised societies. Thus the spiritual and social meanings of water are expressed mainly in abstract terms within the context of religious activities and media, and water is more generally presented as either part of nature (with a focus on the scientific and technical protection of ecosystems) or as a material resource: an economic commodity fundamental to processes of production and central to political economy.

There is, in this dualistic frame, a conceptual alienation of humankind from 'nature' with 'the environment' seen as a separate material context to be acted upon directively. Some writers have argued that this is also an intensely masculine approach, which subsumes a more feminine concern with social relations, and thus expresses gender and power relations. A product of large-scale, highly mobile capitalist societies, such a distanced vision is somewhat at odds with the development of deep affective attachments to place (Milton 2002). The latter integrated view is widely seen as more typical of indigenous cosmologies in Australia and elsewhere. It is egalitarian, in that it situates humankind in partnership with the environment, and it is holistic, in that social, spiritual, political and economic issues are bound together through the medium of the landscape.

Humans are thus not conceptually alienated from the places and the material environments that they inhabit, instead enjoying deep affective attachment to these and having a commensurate concern to achieve sustainable modes of production and ensure the long-term well-being of social and ecological systems. In considering these alternative models, and the kinds of human-environmental relationships they produce, a familiar theoretical comparison is Ingold's vision of 'lifeworlds' conceived as 'gazed upon' globes and 'inhabited' spheres (2000). As noted elsewhere (Strang 1997), such contrasting worldviews go hand in hand with quite different environmental beliefs and values.

In considering these differences it is important to begin with the caveat that such comparisons comprise yet another dualism, which like all such reductive devices, does not represent the complexities of either alternative, or the reality that each contains substantial elements of the other. There is, in any case, an entire debate about the veracity of an image of indigenous relationships with the environment as a polarised opposite to those of large industrialised societies. There is no need (and insufficient space) to revisit this debate here, but whether such representation is romantic, or whether it genuinely reflects the realities of more localised and stable modes of environmental engagement, it offers a heuristically useful comparative perspective. Representations of indigenous environmental relations offer an implicit critique of more compartmentalised worldviews and the unsustainable environmental relationships that these permit.

In general, contrasts are drawn between 'Western' and indigenous societies, and there is a growing number of ethnographies which make use of this comparative potential to consider broader global differences in modes of environmental interaction. However, larger industrial societies are by no means homogenous: they contain a diverse spectrum of groups, some of whom are inspired by and strive to emulate the kinds of environmental relationships represented by indigenous people. For these groups, the social meanings encoded in water are as important as its role in maintaining healthy ecosystems. Their activities reflect this different emphasis, focusing on creative activities that express ideas about community and attachment to place, and which celebrate spiritual and aesthetic interactions with water. These priorities are very different from those of more conventional water management groups.

This chapter therefore considers two local water groups: one that is orientated towards well-established 'scientific' modes of catchment management, and another that offers an alternative perspective. It explores their different approaches and the beliefs and values that these express. Both groups interact directly with local waterways through a range of activities and, as the ethnography makes plain, many such activities are highly performative and ritualised, underlining the importance of ritual behaviour and sensory involvement in the expression and affirmation of cultural or sub-cultural values. It is also evident that each group's activities both express and engender sympathy for particular political ideologies, positioning them in wider debates about the ownership and control of water.

On Moggill Creek

The Moggill Creek Catchment Group (MCCG) is a local catchment management group in Brisbane. There are about 36 such groups, working on the many small creeks that run into the Brisbane River. The Moggill Creek group is one of the most successful of these: it attracts high numbers of participants; organises regular, well-attended activities; and has garnered considerable financial support and expert assistance. Like most such groups, it has been established by local actors in collaboration with (and with the support of) various Government agencies. It therefore reflects a typical process in which prominent actors within a local social network come together with local Government experts to consolidate their activities into a more formal organisation which then, through its representations and activities, develops a definable collective identity and serves to articulate the shared aims and values of the group.

Moggill Creek emerges in the western uplands of the city, in an area originally logged for its red cedar and hoop pine in the 1800s. A reminder of this history remains in the Rafting Ground Park, from which logs used to be floated down the river. The catchment was then cleared for agriculture:

There was a lot of fruit growing—we've been here forty years and when we came here there were bananas and pawpaws being grown on a lot of the hillsides, custard apples to some degree. In the valleys there was a big market garden, vegetables and so on, a little bit of dairy farming... So it was quite a farming area. (Bryan Hacker)

Much of the catchment is still a semi-rural area with some sizeable blocks of grazing land and forest, but as agriculture became less profitable and the city was expanding, the process of sub-division and development gathered pace. Many former farms are now smaller 'lifestyle blocks', and a number of the catchment landowners keep horses, as the area hosts one of the biggest annual horse shows in Brisbane. While the clearing and farming caused soil erosion, the more recent population of the catchment by people who often have full-time jobs elsewhere has tended to mean more laissez faire forms of management, and invasive weeds such as asparagus vine, cat's claw and lantana are now a major environmental problem within the catchment:

They [these weeds] are swamping the vegetation. They're just over-running it and killing it and they are no use to the native fauna and they are knocking out the flora on which a lot of the native fauna relies. (Bryan Hacker)

As it gets closer to the river, the creek runs through one of Brisbane's most pleasant and affluent suburbs. The catchment area is therefore largely inhabited by well-educated professional families, who own their own homes, some of which come with sizeable parcels of land. The composition of the MCCG reflects this demographic reality. Its committee contains—or draws on the expertise of—a number of academics and students from the nearby University, whose disciplinary interests are mainly in subjects such as botany, environmental studies, geology and psychology. The MCCG has a formally elected committee: a Chair, Secretary, Treasurer, Public Relations officer and so forth, as well as 13 section leaders, each responsible for a particular length of the creek. Some of the members of the group are also experienced managers, and it conducts its business in a commensurately professional manner.

It holds regular, carefully minuted meetings. It has a lively web site and a database of about 300 members who have varying degrees of involvement in its activities. Having gained funding from bodies such as the National Heritage Trust, the Envirofund and the Brisbane City Council (BCC), it keeps formal accounts. It has a strategic plan, and hopes to establish some sort of 'eco-centre'. The group collaborates closely with the Brisbane City Council and Habitat Brisbane, and maintains a range of links with other catchment groups. It liaises with local businesses, some of whom sponsor its activities, and it maintains its relationship with its local community through newsletters, posters, and other representational forms, for example articles in local newsarticles, signage at sites and so on. As the MCCG Chair explains, the group emerged from a local Council initiative in the late 1990s: 'Many of the catchment groups arose in response to the initiative from the council. So it wasn't driven totally by the local community' (Bryan Hacker).

The local Council devised several catchment plans for the main River catchment and organised a Brisbane Catchments Network to coordinate local catchment groups. More recently, the South-East Queensland regional body was set up to funnel Commonwealth Government funding into local catchment group activities, and the MCCG benefited from this, having for several years a coordinator that they shared with the nearby Brisbane Forest Park. At a local level the MCCG had some long-running antecedents, for example in the Rural Environment Planning Association (REPA), which over the last 30 years had tried to exert some restraining influence on the rapid process of sub-division and development that began to transform the area in the 1970s.

I think we've taken over the certainly more sort of biological type activities that REPA was involved in... REPA felt that there was a need for this sort of work to be done and so was doing it, but then when the catchment group started up... REPA stood back and said, 'Well that's what you're doing. Carry on..'. There's no point in competing... A lot of the catchment membership is also REPA membership. (Bryan and Jenny Hacker)

The composition of the executive and core members of the group reflects these interests, and it contains a high proportion of people with expertise in the natural sciences. The group recruits the members of its committee partly by advertising in its newsletter and via a colourful brochure, but mainly through social contacts. As Bryan Hacker observes: 'When we need a position like a Treasurer or a Secretary, we tend to rely on one or other of us saying, 'I know a person up the valley who could perhaps fill that role'. The group is therefore led by a close social network of like-minded people who know each other as friends and neighbours, or as professional contacts. The MCCG has some major aims, some of which echo the responsibilities of local Government agencies:

Caring for biodiversity and caring for land and caring for water, and participation—publicity and participation... It's a very standard set of themes which runs through catchment activities throughout the area... But certainly in caring for land, a lot of the objectives we identified are ones that are being looked after to a considerable degree by REPA... We're very strong in the biodiversity area and we're strong in the publicity area. But caring for land and caring for water, I'm afraid we just don't have the people to do it properly... There's a feeling that we should be doing more water monitoring. The Environmental Protection Agency is no longer monitoring water quality in Moggill Creek. (Bryan Hacker)

But the group also recognises that, in a catchment area of just under 60 square kilometres, there are limits to its abilities to provide sufficient labour, skills and technology for regular scientific monitoring and management over the long term, or to protect riparian areas in which major developments have radically changed the flow patterns of the creek, with major stormwater runoffs and foul flooding (ie. letting sewage into the creek). The MCCG organises a range of different activities: it runs a nursery which provides flora not just for the public areas of the creek but also, in the common interest, to 150 or so riparian landholders. In the last couple of years it has provided them with about ten thousand seedlings, and there is a team of approximately a dozen people who come in several times a month to pot up seeds. The nursery provides an important social centre:

The nursery has been running virtually since we started. It's on Brisbane Forest Park land and they built the nursery and a couple of little offices there and put on the watering system and so on.... It's an extremely good forum for a lot of people who can't help in any other way: there are quite a lot of people who are ageing a bit, who don't feel they can swing a mattock any more. They're very happy to come around once or twice a month and pot up our seedlings and they talk about what birds they've seen on their property and whatever the issues may be, and it's like a good bonding exercise.... It's become, from what I've heard, quite a social event and people enjoy it very much. (Bryan and Jenny Hacker)

The catchment group also conducts surveys of plants and wildlife and holds training workshops. It has a stall each year at the Brookfield Show, and runs photography competitions which are exhibited at the local shopping centre. Key issues for the group (as well as weed and feral animal invasion) include water quality issues; wildlife conservation and concerns about mountain biking and other activities creating erosion and disturbance in the catchment area. It publishes material in its newsletter about related groups and activities, and liaises with a range of groups: for example Natural Resources Management South East Queensland regional group; the Healthy Waterways Partnership and the Environmental Protection Agency. The most regular activities are the group's working bees, usually held on the last Sunday of each month, in which its section leaders and other key players encourage local volunteers to clear non-native plants and weeds from the public areas of the creek and plant native trees and shrubs. The group coordinator reported some typical events to the AGM:

The Brookfield Country Market on 16th October was a busy and fruitful day. The MCCG display was erected and the butterfly plants (Pararistolochia praevenosa) were completely sold out... Half a dozen new members and one renewal signed up on the day and volunteers Don Sands, Dale Borgelt, Gaynor Johnson and myself were flat out keeping up with inquiries!

Weedbuster Week also went very well... Moons Reserve (Section 1) had a busy Weedbuster event on 17th October with approximately 26 people turning up for the Moon's Reserve's first working bee. Thank you to Andrew Wilson and Don Sands for coming along to provide technical instruction and to Judy Walker for networking with the locals and gaining community involvement. Happily a Section Leader has now been found to keep the momentum going at this important site...

Michael Humphries (Section 12) also had a successful working bee on 26th October and was supported by approximately 25 volunteers...

Our Lady of the Rosary Primary School held another planting on 11th November. Bryan Hacker and I were in attendance to 'supervise' this organised chaos... As always it was great fun to work with a succession of enthusiastic children ranging from classes 2 to 7 and they were given MCCG stickers and Green Garden Guide Booklets to take home... [The] Summer Newsletter has now been pretty much finalised by Graeme Wilson and will be distributed at the beginning of December. (Chris Hosking, November 2004)

Much of this work is done by eight small bush care groups, supported directly by Habitat Brisbane, who focus on particular public sections of the creek and are sometimes only loosely involved with the MCCG as a whole.

A working bee in Kenmore provides an example: this is an area that has been fairly recently developed. In accord with the other parts of the Moggill catchment, it is composed of large family homes, each surrounded by quite generous leafy gardens. The creek and its riparian areas now constitutes a communal park winding through the new estates: there is a concrete cycle path along its banks, and at regular intervals, areas of green lawn with barbeque facilities, shaded picnic tables, children's playgrounds, and toilets.

The event took place near the local school, and was timed to coincide with a Planet Ark initiative for a tree planting day. It therefore focused on planting (rather than weed clearing) the objective being to dig holes, and plant and water approximately 460 tree seedlings along the edge of the creek.

On this occasion the working bee brought in about 45 volunteers, the Planet Ark event doubling the more usual 20 or so. These were generally local residents as well as some people who found out about the event from the MCCG's web site—mainly families with small children, and older, retired people, but there were also some teenagers, and some young professionals. Most were European-Australians, with a few Asian and African individuals. There were no Aboriginal people.

The event was well supported by local Government agencies and industry. A Councillor, Margaret de Wit, looked in to provide encouragement, and Tom McHugh, a member of Habitat Brisbane, came to provide support. Some of the people carefully digging holes were being sponsored by Ergon Energy on behalf of its 'green team' who, as well as participating in this kind of thing to enhance the company's community outreach, hope to encourage tree planters to make sure that their efforts are placed so as to be 'power line friendly' (Blondine Cousin).

The group's activities were carefully coordinated and somewhat ritualised. The event began in the early morning, with a view to getting the job done while the day was still cool. As people arrived each was asked to sign an insurance form. They were then issued with mattocks, gloves and spades, and directed by the section leader, Malcolm Frost, towards specific tasks.

The event had required forward preparation. Tools had been brought, hosepipes had been connected to nearby standpipes, and buckets provided to carry water up from these or from the creek to water the seedlings as they went in. The Council had supplied a large amount of mulch to protect the new trees. Cold water was provided for the workers throughout the morning, and snacks and drinks had been brought for the conclusion of the event. Some families had also brought picnics. These refreshments were welcome by late morning, when all 460 trees had been planted.

The Moggill Creek group has been active for some years, with the result that, in planting this section of the creek, longer-term participants were also able to inspect sections a little way upstream, which they had planted in previous years, and to chart the progress of each 'year' of trees. They were also able to discern changes in the creek, commenting that it is much lower now than it was a decade ago ('when this was only green fields'), because of land development, new dams and landscaping, and the recent years of drought. Previously, the water was deep enough for children to ride horses into the waterholes to swim: now the water levels are very low, and the water quality is more dubious.

Nevertheless, it was plain that, while some of the problems in the creek are the direct result of the development that has enlarged the local population, the group lavishes considerable care on the creek. This exercise has considerable meaning for them: quite young children recalled which trees they had planted in previous years, and one mother commented that the kids had all wanted to have their names on the trees they planted.

The Planet Ark link brought in a number of new people for this particular tree planting, but it was evident that the major actors in the group were well known to each other. The MCCG's Publicity Officer, Christine Hosking, describes the group as being composed of a series of concentric circles: an 'inner circle' of very committed individuals; a larger circle of 'doers'; a wider one of people who are sometimes involved but 'come in and out more'; and one of casual, occasional participation. She suggested that, like a number of others, she had joined the group when her children left home, feeling a need to do something worthwhile, and this had led to a course in landscaping. As Margaret de Wit observed, social networking is a central part of these events, or as Tom McHugh put it, 'this is how they get to know their neighbours'. At the time of writing, the group was conducting its own research into people's motivations for participating, with the help of a psychologist, Brad Jorgensen at the University of Queensland. Bryan and Jenny Hacker describe a range of motivations for the group: Some of them do so to gain power—they like controlling things. Some do it to gain the experience and they're just in it for a couple of years, and they can use that when they are applying for a job. In our case, I suspect the majority do it because they like to feel by contributing they are making our area a better place. Some might do it because they're frightened of Malcolm! (Bryan Hacker)

They described their own motivations too: I think a sense of duty as far as I'm concerned. I care about it, you know. When I took my dad around two years back, he was saying how lucky we were that there weren't little houses all over the place. I was thinking, 'Well, maybe the fights we've had are worthwhile'... The environment—stewardship... It's not necessarily your social contacts—you have a knowledge, therefore you feel you shouldn't be squandering it... I think it's more concern for the environment of the area. For many people the driving force is money, or fast cars or sport, or whatever. They don't give a damn about whether something is a Chinese elm or a lantana... (Bryan and Jenny Hacker)

There is also a strong element of problem solving, meeting challenges and recreating a previous ecological order:

We started off with the concept that we would like to restore the water weeds to the condition that they were in, [or] as close as possible to the condition they were in two hundred years ago... Now, how one goes about this is very difficult. Where the vegetation is cleared away from around waterways, there's too much light, the water is a higher temperature—you know all this stuff... I think most of us Greenies have the idea that we'd like to restore the vegetation along our bit of creek. (Bryan Hacker) Thus the group has lively debates about whether it should rely on purely local genetic stock (which in some cases is very limited) or bring in new stock from further afield. However, it achieves a broad consensus on what are regarded as weeds, while accepting that some of the exotic plants brought in for purely decorative purposes, such as the omnipresent jacarandas, have aesthetic value to many people.

> We do find people who are leaving camphor laurels because they are pretty trees... when I'm talking to landowners, I can say, 'Well I can't criticise you because I've got two big jacarandas out here and they're a weed species too,' and I wouldn't chop them down for anything. So you can't be didactic. (Bryan Hacker)

Through its caretaking rituals and wider activities, the MCCG has succeeded in generating a collective sense of ownership and responsibility in relation to a shared cultural waterscape, and in doing so, contributed to the social cohesion of a relatively new local community, enabling its members to identify with their local area and with like-minded neighbours. It is greatly enabled in this by the reality that many of its members also own small patches of the catchment area, and are thus able to act directly upon them. Although putatively directed towards conventional ecological management, it supports social needs as well, providing a way in which local residents can meet and act collectively, and construct both social and environmental relationships.

These relationships are further linked with wider social networks and larger agencies involved in environmental management. The group's social network provides close ties with the institutions of governance and education, and it is empowered by considerable support from these. In turn, it engages with the catchment area in a way that effectively extends the ability of local government organisations to manage the cultural landscape and the waterways of the Brisbane area.

In the activities of the group it is possible to see, underlying it, an implicit vision of particular social order. Many of the residents in the catchment have been there for many decades, and are much attached to the area. They have a clear idea of how the catchment should be populated and managed, expressed for example in the way that MCCG recruits its executive officers, and in resistance to the pollution of incomers in excessive numbers despoiling the catchment with 'little houses all over the place', and in the emotive force of the group's passion for re-ordering the local ecology. The MCCG's activities also enable a direct physical engagement with the creek and its environs. There is sensory pleasure in digging holes in the mud and pouring water in them, in handling the delicate tree seedlings and scattering sweet smelling mulch around them. For people mainly confined to office work, the opportunity to physically exert themselves in a common task is rewarding.

Nevertheless, the discourses of the group are heavily dominated by scientific and ecological concerns, and the meanings that are foregrounded in this conceptual model present water as the basis for ecological rather than social regeneration. In this respect the MCCG conforms to mainstream patterns of water resource management, valorising particular knowledges and values, and enacting these in material terms. Moggill Creek is thus the object of an explicit shared scientific and managerial task, and the basis of a more implicit form of social cohesion.

Making a Splash

A less conventional mode of engagement is expressed by the Splash! Festival in Maroochydote, on the 'Sunshine Coast' north of Brisbane. Splash! has taken place every two years since 1999, and is organised by a partnership between Maroochy Water Watch and a Festival design team. The latter is composed of an artistic director, a community co-ordinator, a dance co-ordinator and a music co-ordinator, and several local artists, including an indigenous artist. This team provides links with the wider social network of music groups and artists who perform or exhibit at the Festival (as well as in local exhibitions and related events).

Demographically, the Festival brings together a very different social mix from that found in the Moggill Creek catchment area. It is younger, less academically inclined, and in general less well-off. Most of those centrally involved could be loosely described as members of the 'alternative' community, which tends to place them in the lower economic brackets and on the political left. Many are urban dwellers and live in rented accommodation, and thus have little direct ownership of the Maroochy River catchment area. In some cases their domestic arrangements focus around 'alternative' lifestyle block ownership and communal living experiments of various kinds. The catchment itself is about an hour's drive from Brisbane and is located in an attractive rural area, highlighted by the dramatic landscapes of the Crystal Mountains and Maleny. It has long been popular with 'blockies', and there is a high number of New Age 'retreats' and spas, cottage craft enterprises, nature parks, small organic horticultural businesses and hobby farms. Though 'alternative' in its demographics and activities, this is fundamentally a rural community in which people know each other through participation in local events and everyday activities. The Festival Committee is therefore readily able to disseminate information about its activities by word of mouth, and (like the MCCG), supplements these with a newsletter, advertising and a web site. In the range of people involved in the Festival, there is some overlap with the membership of local catchment care groups or wildlife conservation organisations. Thus the Splash! Festival provides stalls enabling a number of these groups to advertise their activities.

Other than this, however, it has little engagement with the kind of scientific, managerial perspective that dominates the MCCG and other catchment groups. Instead it is deliberately directed towards what could be considered as the opposite perspective: one that focuses entirely on social, spiritual and aesthetic interactions with the local waterways. In its own way this is as one-sided as its polar opposite, reflecting a conventional art-science, female-male dichotomy, but like the Moggill Creek Catchment Group, it contains implicit elements of its oppositional 'other' in its attempts to encourage change in environmental practices. For the Splash! Festival, though, the route to this is via collective social change, rather than through the managerial application of science. By conflating concepts of community and ecological well-being, and presenting the local waterways as a 'connective element', the purpose of the Festival is to challenge and subvert what it perceives as an exploitative and materialistic use of land and resources and encourage people to embrace more holistic and sustainable forms of interaction with the environment. The Festival organisers describe it as:

> *... a unique educational resource connecting people to the waterways environments and thus towards the solutions for our environment. A community based cultural celebration of the Sunshine Coast waterways and people from the range to the sea... Splash! is an opportunity for communities to creatively explore our relationship to the waterways. And in doing so, be aware of how we are connected to other communities via these waterways. (Kingfisher Connection, Bulletin 11. 2004:1)*

The Festival actually begins some time before the main event. There is a range of workshops over the preceding weeks, in which the artists involved help schools, colleges, arts organisations and community groups, as well as local councils and businesses, to create the artworks and performances that constitute Splash! These are all exhibited or performed during an intensive one-day Festival that combines music, dance, theatre, storytelling, participatory sculpture and various forms of 'play' with hydrolarrous rituals invented specifically for the event.

The Festival organisers have been able to garner financial support from various agencies, either on the basis of the event's 'artistic' contribution, or in the hope that it will assist the efforts of more managerial agencies to achieve greater environmental and social sustainability, for example by persuading people to conserve water or participate in river catchment care.

It is therefore included as a National Water Week event, and funded by several bodies including Arts Queensland, Festivals Australia, Maroochy Shire Council and Healthy Waterways.

It is further supported by the International Volunteers for Peace; by large environmental organisations, such as Green Corps; and by a number of smaller local catchment groups. In the direct organisation of the event, Maroochy Water Watch works in partnership with the Festival design team to coordinate the participating groups and their various contributions. As one of the pre-Festival newsletters put it:

Splash! is like a jigsaw puzzle. Each group creates a piece/a chapter. Together we will tell the story of the Sunshine Coast waterways. (Kingfisher Connection. Bulletin 11. 2004:1)

The Festival takes place on Chambers Island, which is an islet (little more than a sand spit really) in the Maroochy River. Though connected to the town esplanade by a pedestrian bridge, the islet usefully provides a context spatially separated from 'mainland' life, and situated quite literally 'in the river'. The Festival has a 'totem', which in 2004 was the green frog, chosen because this species is a key ecological health indicator. Thus lanterns, t-shirts and people were liberally sprinkled with green footprints, whose purpose is explained as follows: Footprints; where we go, what we leave behind. The frog: an environmental indicator. Follow; learn, look, watch, follow the lead, connect people, land and water. What message do we receive? (Splash! 2004:1).

Participants paid $2 for access to the bridge, and the image of a green frog was stamped on their wrists to provide an entry ticket to the event and identify them with it. Crossing the water, they were met by the banners, posters and stalls of the various groups involved: there were large paintings of different river catchments; stalls presenting information about local endangered species or recruiting for green groups; some arts and crafts exhibitions; organic food stalls and, for later on, a 'South Sea Island Hangi' [sand oven]. As the Festival fliers put it:

Welcome all from the land and the island. The rivers have called us. The creeks, tributaries and estuaries have too. (Splash! 2004:1)

Scattered across the island was a variety of art exhibits focused on aquatic and environmental themes. Most had been commissioned especially for the event, and in 2004 included an anthropomorphic 'tree being'; an aquatic 'house', frog silhouettes, and papier mache water symbols. Even the public toilets were decorated with an aquatic mural. Other artworks, such as a 'footprint mandala' and a frog 'sand sculpture', were constructed on the day itself. Creative activities were organised throughout the day and many of these were for children.

They included the making of butterflies and organic boats, and the face painting without which such an event would be incomplete. A mini-stage/sculpture was created for a children's theatre group, which performed a heartfelt play describing the destruction of 'Spring Town' by the evils of over-development. Also wandering around was a street theatre group, dressed as eccentric 'detectives', whose aim was to elicit help from the visitors in 'finding the footprints of the frog'. The mainstay of the Festival was a series of performances from musical groups, school dance and theatre groups, and a group of Shire Council employees, who mimed a tree-planting event. The audiences were of all ages, with an emphasis on young families, and they meandered informally from one performance to another. Later in the day, the events coalesced. The bands formed up and began to move around the island, their paths deliberately echoing the winding flow of water. One group, its members dressed as water nymphs, played flutes made of water pipes. Another crossed the bridge to the mainland to lead a carnival 'parade' back across to the island. The parade included various totemic animals; the totemic frog; a birdwing butterfly (which is an endangered species), and a large duck-billed platypus. As it flowed off the bridge, the organisers instructed sections of the audience to follow these animals and the musicians weaving like pied pipers around the island. Meanwhile canoes carrying black swan sculptures and Aboriginal dancers from the local Gubbi Gubbi community were paddled across from the far shore. The water ceremony invites us all into a circle. The Gubbi-Gubbi dance their ways of following nature's example. (Splash! flier 2004:1)

Eventually the various flows of people were brought together at the northern end of the island for the high point of the day, which occurred at sunset.

This was the ceremonial 'Pouring of the Waters', in which, accompanied by a generic blessing, water from each of the local rivers (contained in a variety of vessels, all made from 'natural' materials) was poured into a large central vessel to express the social and ecological connections between the communities living in each catchment area. After dark, further musical events supported the 'Butterfly Ball', which was simultaneously a costume party and a general 'end of Festival' social gathering.

Through these activities the Splash! Festival sketches a cultural landscape in which the social and spiritual meanings encoded in water are placed firmly in the foreground. It openly celebrates the idea of water as the essence of social identity and communitas. In doing so, it draws uninhibitedly upon several cosmological traditions: indigenous beliefs about water as the source of ancestral regenerative powers; Celtic and neo-Pagan animism in which water is seen as containing sentient deities; Christian visions of water as of the essence of the holy spirit and social being; and ecological ideas in which water is framed more pragmatically as a 'source of life'.

As well as referring explicitly to these meanings, the Festival takes this a step further, by ensuring that almost all of its events are actively created by local community groups, and in carrying the semi-religiosity of Art to its logical extreme, which is the invention of transformative religious rituals. The participants are encouraged to 'play' creatively and hedonistically. They take part in emotive ritual behaviour, and identify with the water itself. Thus they become—in the true sense of the word—a 'congregation', participating in collective aesthetic, emotional and sensory experiences that challenge their inculcated knowledge and ways of being in the world.

In this way the Festival relies on classic methods of 'enlightenment', using rituals, sound, poetry, movement and visual stimuli to shift people out of their everyday modes of thinking and action, into the adoption and embodiment of new ideas and values. Under the rubric of 'artistic licence' it appropriates and subverts several models of social and environmental interaction, creatively amalgamating ideas about indigenous, neo-Pagan, New Age and 'green' relationships with place, in order to proselytise a new paradigm and encourage the adoption of 'alternative' values.

This vision was encapsulated by a number of the people interviewed on the island, for example Juanita, who described herself, on a beautifully decorative business card, as a 'Messenger of Light'. 'Many people need awakening', she said and 'those of us who are awake are awakening others'. This appropriation and recontextualisation of ideas, and the bricolage of religious and secular concepts, is crucial in providing the moral underpinning for the major themes of the Festival: a determination to affirm the social aspects of engagements with water, and the importance of community, and a related plea for more sustainable forms of environmental interaction. In this sense it challenges the technical model of environmental management that dominates the activities of government agencies and conventional catchment management groups, proposing instead that sustainability will only come through a resanctification of the environment, and greater valorisation of its non-human elements, and the re-establishment of community ties and long-term, affective relationships with place.

This is not a new message of course: Friends of the Earth made it clear in the 1960s and 70s that the environmental movement was equally concerned with social issues. However, although FOE and other long-standing activist groups have continued to foreground the need for social change and a more holistic approach to human-environmental engagement, many contemporary 'green' groups have shifted towards a more specialised—and far less politically challenging—'scientific' focus on ecology and resource management that sits quite comfortably in the context of an economic system still firmly wedded to continual growth and expansion. This shift is clearly reflective of a political reality dominated by economic rationalism and right-wing ideology. This dominance is plainly reflected in public policy that, though it is willing to support uncontroversial economic and scientific approaches to environmental 'services', is intrinsically exclusive of wider social and ecological concerns.

The Splash! Festival can therefore be seen, in some ways, as an attempt to foreground and thus to revitalise the fundamental principles of the feminist and environmental movements that demanded equality both for people, and for 'nature', and which make use of more holistic models of human-environmental relationships, in which social and ecological health are seen as mutually constituted.

This points to another innovative aspect of the Festival which is that, rather than simply presenting a loose romantic vision of indigenous beliefs and values, as often happens in New Age events, the Festival relies upon the active participation of the local indigenous community, and indigenous artists play an important leadership role in its design. This has been particularly influential in establishing a set of practices that valorise co-identification between people and aspects of their local environments.

Thus there is the communal production of material culture that celebrates totemic species, and performances in which people are imaginatively transformed into aquatic animals, butterflies, and water itself.

As has been made clear by a number of writers on environmental issues, this co-identification with place and with aspects of the local environment is crucial to the generation of deep affective concern for its well-being. In this sense, the involvement of indigenous participants has helped the groups concerned to articulate their ideas and develop a poetic language with which to describe connective social and environmental relationships.

There are some undeniable tensions in this respect. On the one hand the Festival sets up a partnership with the local indigenous community and offers a subtextual critique of colonial hegemony and its social and environmental outcomes, but on the other, it also serves to express—and aims to strengthen—the affective attachment and 'belonging in place' of the non-indigenous communities in the area. This could be interpreted as an alignment between 'alternative' minorities, or—as the rather caustic remarks of an Aboriginal spokesperson from a neighbouring community made plain—less benignly, as a further appropriation of indigenous knowledge and land. In a political context where special treatment for indigenous people is being lambasted by right-wing groups as undemocratic and discriminatory, the inclusiveness of the Festival's vision of a community of belonging, and its efforts to express the equality of all humankind, may be somewhat of a double-edged sword.

Much of the support that the Festival garners comes from the fact that its organisers have been astute in presenting their social and environmental 'message' under the apparently benign umbrella of an Arts Festival.

They are not the first to rely upon the protective mystification of art to disseminate a political critique of dominant paradigms: there is a well established tradition of this in the Arts, and organisations such as Common Ground in the UK and others have long recognised the utility of engendering social and environmental action through the licence provided by artistic events.

Another example is the 'Tipping Point' conferences that take place annually in the UK, bringing climate change scientists and artists together in a creative collaboration. And Brisbane has its own (very large) annual River Festival, which also aims to bring science and art together in a productive way.

Through their engagements with water, both the Splash! Festival and the MCCG work to construct and maintain local communities, but the social networks that they are attempting to strengthen and regenerate, and the ideologies that they support are quite different from each other.

Each has its own discourses, forms of knowledge, inculcatory ritual activities and values. Each can be broadly aligned with competing conceptual models of 'the environment' and political positions, one of which hopes to critique and subvert a particular status quo, and the other which acts to uphold it.

What is clear is that the Festival promulgates a vision of human-environmental engagement that differs substantially from the forms of knowledge and the values expressed by the Moggill Creek catchment management group. Underlying the jolly music and the totemic parade of the Splash! Festival is a recurrent unease: a sense that a primarily managerial, directive view of human-environmental relationships is too narrow and excludes things that matter.

However, although Splash! owes much to the influence of intensely local and holistic indigenous worldviews, and offers a vision of social and aesthetic rather than ecological order, it remains ineluctably an abstract Western vision in which 'the environment' is a somewhat distanced object.

In this sense it confirms Argyrou's assertion that contemporary environmentalism remains a 'modernist project' (1995). The Festival is an ephemeral event, and it is difficult to say what the outcomes would be if it was accompanied by similar activities on a regular basis, or if the participants were empowered by land ownership, economic strength or links with Government bodies to act upon the waterways more directly.

As it stands, it may have rather less capacity for community building than the Moggill Creek group, in which long-term social relationships are constructed and local land and property owners are regularly brought together in collective enterprise that involves direct management of the material environment. It is plain that, empowered by its political and social relationships with governing elites, educational institutions and industries, and by its reliance upon scientific authority, the MCCG has a greater capacity to act upon the local environment, to influence wider regional management and to affect the development of public policy.

The Splash! Festival, while it benefits from some linkage with local government bodies and sponsors, has few powerful political alliances and only loose connections with groups underpinned by 'science'.

Its 'artistic' nature, while it serves to obscure the political content of its message, also means that it can be readily marginalised, and excluded from the processes through which real managerial decisions are made about Queensland's river catchments. The Festival's alliance with indigenous people and their beliefs and values, while inspirational, is unlikely to be empowering in the contemporary political climate in Australia. What Splash! achieves, however, is an explicit commentary about the social aspects of people's engagements with water, which is generally absent from more focused scientific discourses, and which does exert its own influence upon local debates.

It serves to foreground a critique of 'Western' engagements with the environment, mentioned at the beginning of this chapter that, having subsided somewhat since its emergence in the 1960s, is undergoing a resurgence as the realities of climate change and chronic water shortages hit home. This small comparison therefore brings to the surface some of the social, political and intellectual divisions that underlie contests over water in Australia, and which are manifested in many of its river catchments.

An understanding of these differences is useful for policy makers and water managers whose energies are directed towards encouraging community involvement in catchment management, and resolving conflicts between interest groups. The study makes an obvious case for combining the strengths of these different kinds of approaches, bringing social issues to the foreground for 'management' groups, and providing more 'socially and artistically' inclined groups with greater potential to act upon their ideological aspirations. Such integration might enable some reconciliation of viewpoints that, at present, occupy opposing positions.

4

Separate Storm Sewer Systems: Water Quality Standards

In 1987, Congress added section 402(p) to the Clean Water Act, which addresses point source storm water pollution, including discharges from municipal separate storm sewer systems (MS4s). The MS4 program, regulations for which were promulgated by EPA in 1990 and 1999, is just beginning to show some progress in improving storm water quality. However, EPA's program has not required MS4s to meet measurable standards or goals, or ensured compliance with water quality standards. In 1999, the Ninth Circuit held, in Defenders of Wildlife v. Browner, that the Clean Water Act does not require MS4 permits to ensure compliance with state water quality standards. That holding, if followed in other circuits, could stunt the progress of the MS4 program.

This Chapter suggests that the weaknesses in the MS4 program—particularly the lack of storm water quality monitoring and characterization and lack of required measurable standards or goals—must be addressed for the program to be ultimately effective. Finally, this Chapter concludes that EPA should require MS4s to comply with water quality standards, a goal that can be more effectively achieved by requiring minimum performance standards and by incorporating total maximum daily loads (TMDLs) for water quality-limited waters into MS4 permits. Storm water runoff is a leading cause of water quality impairment in many water bodies of the United States. In 1998, the Environmental Protection Agency (EPA) reported that, of the waterways assessed by the states, urban runoff, or municipal separate storm sewer system (MS4) discharges, are a significant pollutant source affecting impaired waterways. There are many examples across the nation of the severity of the problem.

In the Pacific Northwest, endangered salmon are adversely affected by sediment, nutrients, and toxic chemicals found in urban storm water. In 1998, over 7000 days of beach closures nationwide occurred due to poor water quality, for which the main culprits were elevated bacteria levels and specific pollutants that reached waters from sewage spills and urban storm water, In 1997, excessive nutrient pollution in storm water runoff caused a toxic algae bloom in Maryland that left at least thirty people ill and caused fish kills and contamination resulting in $43 million in fishery losses.

In spite of these occurrences, the 1999 Ninth Circuit ruling in Defenders of Wildlife v. Browner that MS4 permits need not require more stringent limitations necessary to comply with state water quality standards—if followed in other circuits—may allow MS4 discharges to continue to threaten the health of the nation's waterways. Storm water is a threat to water quality because as rainfall or snowmelt moves over the land it picks up numerous natural and human-made pollutants that are discharged directly into waterways. Urbanization multiplies the adverse effects of storm water. Development creates large impervious areas—such as rooftops, roads, parking lots, and sidewalks—which change the hydrologic cycle by preventing storm water from infiltrating the soft.

Reduced infiltration impedes the natural filtering and removal of pollutants and the recharge of groundwater, which results in storm water flows that are higher in speed, volume, and temperature. A great variety and number of pollutants reach waterways through urban storm water and as population density increases, so does the concentration of pollutants. These adverse water quality and quantity effects combine to increase pollutant loads, degrade or destroy aquatic and riparian habitat, and reduce the number and diversity of fish and macroinvertebrate species. Storm water discharge threatens human health by increasing pollutant loads, such as bacteria, in water supplies and recreational areas, and degrades aesthetic conditions by discharging floatable trash, oil, and grease that cause surface scum and odor. The effects of polluted urban storm water are preventing many waterways from achieving acceptable levels of water quality—not the least of which is the federal Clean Water Act's (CWA) "fishable/swimmable" goal.

The CWA seeks "to restore and maintain the chemical, physical, and biological integrity of the Nation's waters" by controlling and abating discharges of pollutants into the waters of the United States.

The National Pollution Discharge Elimination System (NPDES), under section 402 of the CWA, requires entities seeking to discharge pollutants into a waterway to obtain an NPDES permit, subject to conditions set by the permitting authority. The CWA also requires each state to establish water quality standards for its waterways. Generally, NPDES permits must contain additional conditions designed to meet those standards if the applicable NPDES technology standard is not reducing the discharge of pollutants to acceptable levels. Therefore, the CWA requires an MS4, as a discharging entity, to obtain and comply with an NPDES permit.

MS4s capture and discharge storm water from outfalls into the waters of the United States. These outfalls are considered point sources under the CWA. Due to the significant adverse effects of storm water on water quality, MS4 discharges are a significant target for pollution control. However, MS4 discharges are not as amenable to traditional pollution control measures as typically targeted water pollution sources, such as industrial discharges. Storm water flows are intermittent and unpredictable because they fluctuate depending upon frequency and duration of precipitation. Further, MS4s generally contain numerous outfalls (so designed to minimize flooding), the location of which may not be completely known to the municipality.

These problems can make sampling and measuring the water quality impact of storm water difficult and in some cases, impractical. Moreover, local governments' limited financial resources can make traditional end-of-pipe treatment infeasible for many municipalities. Despite these difficulties, MS4s require regulation to preserve and improve the quality of the nation's waters. In part to specifically address the regulation of storm water discharges, Congress passed the Water Quality Act of 1987. The Act extensively amended the CWA, including the addition of section 402(p), which addresses storm water discharges. The amendment directs EPA to proceed in two phases under the NPDES program. The Phase I regulations, promulgated in 1990, require permits for storm water discharges from industrial sources, construction disturbing over five acres, and large and medium MS4s. In 1999, the Phase II regulations were promulgated, adding small construction sites and small MS4s into the fold of regulated storm water discharges.

EPA attempted to create a flexible program to address MS4 pollution because of the inherent difficulty in controlling storm water and the sensitive nature of regulating municipalities. The regulations outline a program that is substantially different from the other point source water pollution control programs under the CWA. The MS4 storm water program uses a "performance-based" regulatory approach instead of the traditional "technology-based" approach used in end-of-pipe treatment. Whereas a strictly "technology-based" approach focuses on attaining set effluent limits through the use of an end-of-pipe technology control, the "performance-based" approach focuses on the implementation of an overall plan that employs various methods of pollution control. Municipalities are required to develop storm water management programs (SWMPs) that include best management practices (BMPs) for reducing storm water pollution; these plans become permit conditions. This approach allows for municipal flexibility in tailoring programs for particular circumstances.

In 1999, environmental groups brought the first lawsuit challenging the content of an MS4 permit. In Defenders of Wildlife v. Browner, Defenders of Wildlife sought review of an EPA decision to issue MS4 NPDES permits for five municipalities in Arizona. Defenders of Wildlife argued that EPA violated section 402(p) of the CWA because the permits did not contain conditions requiring strict compliance with state water quality standards through the use of numeric effluent limits. The Ninth Circuit disagreed, determining that section 402(p)(3) does not require strict compliance for MS4 permits. However, the court determined that EPA, in its discretion, may require compliance with state water quality standards and can determine the level of, and method for, compliance.

There is limited information on whether EPA's approach to MS4 regulation is actually resulting in storm water discharges that comply with state water quality standards. The studies that have been conducted focus on particular BMPs' ability to remove pollutants or control storm water volume and velocity. These studies indicate that the implementation of appropriate BMPs does result in improvements in storm water quality on a site-specific basis, but information about watershed-scale improvement is lacking. Further, EPA's flexible regulatory approach provides little guidance or standardization for monitoring BMP effectiveness, nor has EPA required MS4s to meet measurable goals.

This Chapter examines the MS4 storm water program, focusing on the potential for attainment of state water quality standards. This Chapter concludes that the MS4 storm water program has made a slow, but encouraging beginning; however, if MS4 attainment of water quality standards is a goal of the states, future permits should require attainment of minimum standards and goals by which SWMP effectiveness can be measured and enforced, such as integrating total maximum daily loads (TMDLs) into MS4 permits.

History of the Storm Water Programme

1972-1987: Lawsuits and Proposed Rules Without Results

Clean Water Act Gets Teeth : In 1972, Congress significantly amended the Federal Water Pollution Control Act—commonly referred to as the Clean Water Act (CWA)—"to restore and maintain the chemical, physical, and biological integrity of the Nation's waters." Among the most important additions was the NPDES program, which requires permits for discharges of pollutants into navigable waters from point sources. Under this program, entities seeking to discharge pollutants into a waterway must obtain an NPDES permit. Permits contain conditions set by the permitting authority, including a technology-based standard that establishes limitations on the amount of pollutants that may be discharged. Once the NPDES permit is issued, the discharging entity satisfies the requirements of the CWA as long as the permit conditions are not violated.

The CWA also requires states to establish water quality standards for their waterways. These standards can be expressed as narrative conditions or numeric effluent limits for particular pollutants. Under section 303 of the CWA, the state must establish TMDLs for water bodies that fail to meet water quality standards (water quality-limited waters). TMDLs are estimated gross amounts of pollutants that a water body can absorb without causing a water quality standard violation. A TMDL is established by first estimating how much of a particular pollutant the waterway can absorb. Second, the state must apportion the TMDL among the pollution sources in the TMDL area. This process must take into account both point and nonpoint sources of pollutants. Also, a portion of the TMDL must be reserved for a margin of safety, to account for any gaps in knowledge and seasonal variances. Typically, NPDES permits must contain more stringent limitations than the applicable technology-based standard if the discharge is contributing to a violation of state water quality standards; these limitations are based upon TMDLs for water quality-limited waters. Thus, the water quality standards serve as a "safety net," so that the CWA's fishable/swimmable goals can be met even if those goals are beyond the reach of established technology requirements.

NPDES and MS4s

Although storm water runoff begins as a nonpoint source, the collection and discharge of storm water from outfalls into water bodies by MS4s constitutes a discharge from a point source. Since the inception of the program, MS4 outfalls have been subject to the NPDES program, but only after several lawsuits and congressional intervention did MS4s widely begin to obtain NPDES permits.

In 1973, EPA promulgated regulations exempting from the NPDES program MS4 discharges that did not contain industrial or commercial contamination. The D.C. Circuit held in 1977 that EPA could not exempt classes of point sources from the NPDES permit requirement. EPA argued that it should be able to create exemptions because it would be infeasible to establish national effluent limitations for this type of discharge. The court rejected that argument, holding that the NPDES permit program gave EPA the flexibility to issue general or area permits and establish permit conditions that were not strict numeric effluent limits. The court made it clear that EPA was required to regulate MS4 discharges under the NPDES permit program.

In 1980, EPA's first attempt at promulgating regulations for storm water discharges was challenged by both industry and environmental groups. In a proposed rule that reflected a settlement with industry groups, EPA agreed to require permits only for storm sewers discharging effluent contaminated with specific pollutants. Practically ignoring Costle, EPA had agreed to exempt certain storm water point sources from regulation. However, in its final 1984 rule, EPA decided not to narrow the class of covered discharges; but it retained the two-tiered regulatory approach for MS4s, proposed in 1982, that did not require MS4s to control the pollution in their discharges. Again, industry and environmental groups challenged the regulations, but in light of the recently enacted Water Quality Act of 1987, the D.C. Circuit vacated and remanded the regulations upon a request by EPA.

The Water Quality act of 1987

Congressional concern over the serious adverse effects of storm water discharges on water quality and a general frustration with EPA's inability to promulgate satisfactory regulations prompted the addition, in 1987, of section 402(p) to the CWA. Section 402(p) sets up a basic program for storm water discharges that establishes priorities, deadlines, and application requirements. It also provides some relief to municipal and nonindustrial entities from the typical NPDES permit requirements.

Congress first provided a general moratorium on storm water permits until October 1, 1994, subject to five exceptions: 1) discharges already regulated under an NPDES permit, 2) industrial discharges, 3) large MS4s, 4) medium MS4s, and 5) any discharge designated by EPA or the State as a significant contributor to violations of water quality standards or pollutants into a waterway. Congress distinguished between large, medium, and small MS4s in order to place priority on the discharges that were major contributors to the impairment of water quality. This statutory scheme enabled EPA to split up the MS4 regulatory program into two phases. Phase I covers discharges from large and medium MS4s; Phase II covers small MS4s.

Section 402(p)(3)(B) sets out the relaxed permit requirements for MS4s. The storm water program allows for issuance of system-or jurisdiction-wide permits, which alleviates the difficulty of permitting every discharge point, of which a large MS4 may have hundreds. The permits must include a prohibition on non-storm water discharges into the MS4 system and "shall require controls to reduce the discharge of pollutants to the maximum extent practicable, including management practices, control techniques and system design and engineering methods, and such other provisions as the Administrator or the State determines appropriate for the control of such pollutants." "Maximum extent practicable" (MEP), the treatment standard for MS4s, was a new CWA technology standard that Congress did not define. The legislative history indicates that Congress intended that site-specific standards would be used rather than traditional end-of-pipe limits. These relaxed standards enabled EPA to create an MS4 program that is more flexible than the traditional NPDES permit requirements.

Section 402(p)(5) directed EPA to conduct a study designed to better understand storm water discharges and how to control their adverse effects on water quality. The study was supposed to accomplish three goals: identify storm water discharges that fell under the permit moratorium, determine "the nature and extent of pollutants in such discharges," and establish methods and controls for storm water that will protect water quality. Congress directed EPA to use the results of the study to assist in developing the regulations for storm water discharges covered by the moratorium. EPA completed this task with the promulgation of the Phase II regulations.

Regulation Delays

The Water Quality Act of 1987 directed EPA to promulgate rules for large MS4s by February 4, 1989, medium MS4s by February 4, 1991, and small MS4s by October 1, 1993. EPA missed the deadlines for large and small MS4s. Phase I regulations, which cover large and medium MS4s, were not promulgated until November 16, 1990 after much controversy. In response to the proposed rule, EPA received over 3200 pages of comments from affected industries, trade association, municipalities, state and federal agencies, environmental groups, and private citizens. After EPA missed the 1989 deadline, a citizen filed suit to compel promulgation of regulations? As a result of the lawsuit, EPA entered into a consent decree, which required regulations by October 31, 1990. EPA also missed this deadline, but only by a few weeks.

Phase II regulations, which cover small MS4s and small construction sites, took EPA another nine years to develop. The agency conducted extensive public outreach, particularly with small communities, which contributed to the delay. EPA requested comments on three key issues in September 1992 as a starting point for developing the rule. In 1993, EPA conducted several public and expert meetings that included outreach to representatives of small municipalities. The Agency also established the Urban Wet Weather Flows Federal Advisory Committee, composed of representatives of interested parties to advise on the proposed rules. Ultimately, the final rule extended the application deadline for small MS4s to March 10, 2003. The result of this continued delay, which began at the creation of the NPDES scheme in 1972, is a storm water program that is still in its early stages with many discharges remaining unpermitted and uncontrolled.

The MS4 Programme

The permitting approach developed by EPA for MS4s differs from the typical effluent limits used in NPDES permits. First, the program requires MS4s to develop storm water management programs (SWMPs) and use BMPs to reduce pollution, rather than requiring end-of-pipe technology controls. Second, EPA encourages municipalities to use a watershed management approach to address storm water pollution control. To facilitate this approach, EPA allows the use of system-wide or general permits and encourages multiple MS4s within the same watershed to obtain co-permits, This allows different entities that operate storm sewer systems in the same area to combine efforts on their SWMPs that are tailored to the needs of the particular watershed. This permitting approach also precludes the need to obtain a permit for every outfall within an MS4. Third, as provided by section 402(p)(3), the MS4 technology standard reduces the discharge of pollutants to "the maximum extent practicable" (MEP), a standard that both Congress and EPA have declined to define. This approach to pollution management and NPDES permitting has resulted in a flexible regulatory program that leaves maximum control with the municipalities.

Phase I Regulations for Large and Medium MS4s

Phase I regulations were promulgated November 16, 1990 and require all medium and large MS4s to obtain NPDES permits for their storm sewer discharges. EPA and the states have designated 1059 MS4s as subject to Phase I regulations, 1017 of which have received or are in the final stages of receiving an NPDES permit as of February 2000. EPA recently updated the list of large and medium MS4s to reflect the 1990 census and as of promulgation of the Phase II regulations, has frozen that list. All other MS4s must proceed under Phase II regulations, even if in the future the MS4 grows to technically qualify as large or medium.

Phase I focuses upon the application process, which is divided into two parts over two years and requires the "development and implementation of a local, site-specific [SWMP]." SWMPs must include a plan to: identify major outfalls and pollutant loads; detect and eliminate non-storm water discharges; reduce pollutants from industrial, commercial, and residential areas; control discharge from new development and redevelopment; and meet the treatment standard of "reducing pollutants to the maximum extent practicable."

Part one of the application primarily focuses on gathering information about the MS4; it requires the municipality to describe any existing programs, the means available to it to control pollutant discharges, and information on a field screening program to identify illicit connections. A year after submitting part one, the MS4 must complete part two—which expands upon the information and the SWMP initially provided in part one—based upon representative data gathered from major outfalls. Part two of the application has six elements, requiring the MS4 to 1) demonstrate that it has adequate legal authority to implement the SWMP, 2) describe the location of major outfalls and identify industrial facilities that discharge storm water through the MS4, 3) provide representative data in accord with an approved sampling plan, 4) describe the proposed SWMP, 5) describe assessment measures to determine the effectiveness of the proposed SWMP, and 6) provide a fiscal analysis of necessary expenditures to operate and maintain the SWMP. In addition, when the permit covers more than one legal entity, the application must address the individual responsibilities of each legal entity and how pollution control efforts will be coordinated.

Once the application is complete, the permitting authority issues an NPDES permit that incorporates the SWMP. The permit conditions consist of implementation of the program and any other requirements to ensure proper management, such as a timetable for implementation. Also, the MS4 must submit annual reports that include a status report on SWMP implementation; any proposed changes to the SWMP; a summary of data collected in the previous year; annual expenditures and the budget for the coming year; a summary of enforcement actions, inspections, and public education programs; and identification of water quality improvements or degradation. Sampling data and monitoring focus on major outfalls, with testing limited to 250 outfalls for medium MS4s and 500 outfalls for large MS4s. MS4s use models to estimate pollutant loads and concentration.

Although large and medium MS4s are required to "submit information on source control methods and estimate the annual pollutant load reduction,... they are not required to achieve any specified level of reduction of any pollutants." EPA's failure to provide any definitive measurable requirements for MS4s is the program's weakest aspect. Because MS4s are not required to achieve any measurable goals, whether the MS4 is properly and effectively implementing its SWMP is difficult to determine, at least by objective means.

Phase II Regulations for Small MS4s

Six years after the statutory deadline, the Phase II regulations were promulgated on December 8, 1999. These regulations provide NPDES permit coverage for "donut holes," which are unregulated, smaller MS4s located within urban areas that include a larger Phase I MS4. EPA designated 5040 small MS4s located in "urbanized areas" that will be regulated under Phase II. The state permitting authority must also designate other small MS4s whose discharges "result in or [have] the potential to result in exceedances of water quality standards."

The permitting authority is required to develop a process and criteria to make these designations. EPA suggests using six designation criteria which consider whether the MS4: 1) discharges to sensitive waters, 2) has high growth potential, 3) has a high population density, 4) is contiguous to an "urbanized area," 5) is a significant contributor of pollutants to waters, and 6) is ineffective in protecting water quality through other programs. This requirement mandates that at a minimum, the state authority examine all MS4s serving a jurisdiction with a population density of at least 1000 persons per square mile and a population over 10,000; however, states are encouraged to examine smaller MS4s and designate those that meet the established criteria. Citizens may also petition the permitting authority to designate small MS4s for Phase II regulation.

There are two types of permit waivers potentially available for regulated small MS4s. Small MS4s that serve a population of less than 1000 can obtain a waiver if they do not contribute to the pollutant loading of an interconnected MS4 and regulation is not necessary based on a TMDL. MS4s serving a population of less than 10,000 can obtain a waiver if the state has evaluated all the receiving waters for that MS4 and regulation is not necessary based on an established TMDL that addresses pollutants of concern. In general, a small MS4 is covered by the Phase II program, unless a waiver is obtained. However, because EPA encourages the use of general permits (including state-wide permits) for regulated small MS4s, the application process and SWMP development is far less comprehensive in the Phase II program than under Phase I.

The permit application for Phase II focuses on selecting appropriate BMPs that address the six minimum measures established by EPA and setting goals for implementation of those BMPs. The minimum measures and the selected BMPs become the MS4's SWMP. Each MS4 must develop and implement the following: a public education and outreach program to inform the public about storm water impacts; public involvement or participation programs that include the public in developing, implementing, and reviewing the SWMP; a program to detect and eliminate illicit discharges; requirements (through ordinances and site inspections) for construction sites to control pollutants in storm water runoff to the MS4; a program (using BMPs and ordinances) to address post construction storm water management in new development and redevelopment; and an operation and maintenance program that prevents or reduces pollutant runoff from municipal operations. The Phase II MS4 must also submit annual reports for the first permit term and in years two and four for subsequent permits. Small MS4s are required only to proceed with good faith implementation of their selected BMPs to be in permit compliance. Under Phase II, like Phase I, the MS4 is not required to achieve pollutant load reductions or improvements in ambient water quality. Further, EPA strongly recommends that permitting authorities not place any additional requirements beyond the minimum control measures in Phase II permits until after EPA evaluates the program in 2012. Although EPA professes that the goal of the storm water program is to protect water quality, it does not require that MS4s attain water quality standards, or meet measurable standards for achieving that goal.

EPA's Interim Water Quality Policy

Due to the nature of storm water and the lack of information on which to base numeric effluent limits of pollutant concentration and mass, EPA developed an interim policy for inclusion of water quality limitations in MS4 NPDES permits. The policy applies only to EPA-issued permits, but EPA encourages states with permitting authority to adopt a similar approach. Although this policy was developed prior to the decision in Defenders of Wildlife—which held that MS4 permits are not required to contain any more stringent limitations necessary to attain water quality standards—the policy remains in effect as the preferred method to attain water quality standards. And even though attainment is not mandated (at least in the Ninth Circuit), achievement of water quality standards remains a listed goal of MS4 permits.

EPA's policy is to use BMPs in the first round of permits and in later permits to include expanded and better-tailored BMPs where necessary to attain water quality standards. Where information exists to develop specific conditions or limitations to meet water quality standards, those should be incorporated into permits as necessary and appropriate. Further, each permit should include coordinated, cost-effective monitoring programs. EPA's policy addresses the type of effluent limits appropriate to obtain water quality standards in MS4 permits; it does not pertain to technology-based limits.

This policy reflects EPA's decision not to establish defined standards for MS4s, but rather to improve gradually the type of BMPs an MS4 is implementing in its SWMP as a means of attaining water quality standards. Following the Ninth Circuit ruling in Defenders of Wildlife, it is possible, at least in the Ninth Circuit, that EPA and the states will not even do this much to bring MS4s into attainment with water quality standards. EPA explicitly stated in the Phase II regulations that it strongly recommends not requiring more stringent limitations than the minimum control measures for small MS4s prior to 2012. Permitting authorities may become content with the BMPs implemented in the first round of permits and not require more tailored or stringent controls in later permits, regardless of whether the MS4 is contributing to a violation of a water quality standard.

Ninth Circuit Case Law

Two significant Ninth Circuit cases address the substantive requirements of the MS4 program. In Natural Resources Defense Council v. EPA, NRDC challenged EPA's decision not to define the MS4 technology standard of "reducing pollutants to the maximum extent practicable." In Defenders of Wildlife v. Browner, Defenders of Wildlife sought to require EPA to include numeric effluent limits in MS4 permits in order to attain water quality standards. In both cases, the Ninth Circuit held that the MS4 program as outlined in section 402(p) of the CWA did not require EPA to develop strict standards for MS4 permits.

The MS4 Standard: "To the Maximum Extent Practicable"

In 1992, NRDC challenged several aspects of EPA's 1990 Phase I rule, including a challenge to EPA's decision not to define the MEP technology standard. Prior to the Water Quality Act of 1987, MS4s were subject to the same controls as all point sources, although EPA never successfully promulgated regulations. In section 402(p), the Water Quality Act retained existing controls for industrial discharges, but created a new standard for MS4s: MS4s "shall" prohibit non-storm water discharges into the system and "shall require controls to reduce the discharge of pollutants to the maximum extent practicable."

NRDC claimed that EPA's regulations did not meet the requirements of section 402(p). They argued that the regulations did not achieve the stated goal of effectively prohibiting illicit discharges, that the regulations did not provide minimum controls for meeting MEP as required by the CWA, and that the mandated testing was inadequate because only minimal sampling at a limited number of sites was required. According to NRDC, "Congress granted the moratorium precisely to give EPA the opportunity to develop new, substantive standards for storm water control of municipal sources and instead EPA wrote vague regulations containing no minimum criteria or performance standards." EPA argued that section 402(p) did not mandate the development of a detailed substantive standard, but instead allowed for the effectiveness of individual permits to be decided on a by-permit basis in the discretion of the permitting authority.

Proceeding with a Chevron analysis, the Ninth Circuit, in a cursory opinion, rejected NRDC's arguments. First, the court construed section 402(p) to require a permitting authority to design controls, but stated "Congress did not mandate a minimum standards approach or specify that EPA develop minimal performance requirements." The court concluded that the language of the statute gave the permitting authority much discretion in determining whether an individual applicant's proposed controls are adequate to eliminate illicit discharges and to satisfy MEP. Since the statutory language did not specify controls and gave discretion to EPA, the court, in step two of its Chevron analysis, deferred to EPA's judgment not to mandate minimum performance standards and to determine the adequacy of controls on a by-permit basis.

As to NRDC's third argument, the Ninth Circuit again deferred to EPA's judgment because EPA had "adequately explained" its decision to limit sampling in the application phase. In the publication of the Phase I rule, EPA explained that MS4 discharges are complex and variable, which makes comprehensive sampling difficult. EPA required field screening from representative outfalls to detect high pollutants in dry weather flows, which indicate illicit discharges, and to gain preliminary representative data. The Agency reasoned that the majority of data would be collected during the permit term, and that Congress created this system to "balance the usefulness of [the] data against the economic and logistical constraints in actually obtaining it." The Ninth Circuit's decision left intact EPA's MS4 regulatory approach; an approach that does not define performance standards, does not require attainment of measurable goals, and does not require comprehensive monitoring or storm water quality characterization.

MS4 Compliance with Water Quality Standards

***Defenders of Wildlife v. Browner*:** In 1999, Defenders of Wildlife challenged five municipal permits issued in Arizona, claiming that the CWA requires numeric effluent limitations to ensure strict compliance with state water quality standards. NPDES permits ordinarily impose effluent limitations on discharges under the applicable technology-based standard, mandated under section 301 of the CWA. In addition, section 301(b)(1)(C) requires that a permit holder "shall... achieve... any more stringent limitation, including those necessary to meet water quality standards, treatment standards or schedules of compliance, established pursuant to any State law or regulations." Generally, if the discharge will contribute to a violation of state water quality standards, the permitting authority must impose additional effluent limits designed to ensure that the discharge will no longer contribute to that violation.

Section 402(p)(3) sets out two different storm water standards, one for industrial discharges and one for MS4s. Industrial discharges are required to "meet all applicable provisions of [section 402(p)(3)] and section [301] of this title." MS4s are required to "reduce the discharge of pollutants to the maximum extent practicable, including management practices, control techniques and system design and engineering methods, and such other provisions as the Administrator or the State determines appropriate for the control of such pollutants."

Both EPA and Defenders of Wildlife argued that the standards set out in section 402(p)(3) were ambiguous as to whether Congress intended municipalities to comply strictly with water quality standards, pursuant to section 301(b)(1)(C). EPA and Defenders of Wildlife agreed that the CWA should be interpreted to apply section 301(b)(1)(C) to MS4s; however, they disagreed over the form those effluent limits should take. Conversely, the intervenor cities and amici curiae argued that the provisions were not ambiguous and that the MS4 standard set forth in section 402(p)(3)(B) does not require MS4s to comply with section 301(b)(1)(C).

The Ninth Circuit applied a Chevron analysis to EPA's interpretation of the statute and agreed with the intervenors that the meaning of section 402(p)(3) was not ambiguous, thus, stopping at step one of the analysis. Looking at the language of the statute, the court applied the proposition that "[w]here Congress includes particular language in one section of a statute but omits it in another section of the same Act, it is generally presumed that Congress acts intentionally and purposely in the disparate inclusion or exclusion." Here, Congress expressly required industrial storm water discharges to comply with section 301, which includes strict compliance with water quality standards. However, Congress used different language in the MS4 standard. MS4s must reduce pollutants to the maximum extent practicable, but Congress did not mention section 301 as it did in the industrial standard.

Applying the principle, the Ninth Circuit concluded that the MS4 standard unambiguously does not require strict compliance with section

301(b)(1)(C). The court held that this was the only proper conclusion because otherwise both the statutory sections would not be given effect, a result courts prefer to avoid. Reasoning that the MS4 MEP standard created a lesser requirement than section 301, the court held that the MS4 standard must replace all of section 301, otherwise section 301 would always control, rendering the MS4 standard superfluous.

Finally, the Ninth Circuit supported its interpretation with its decision in NRDC v. EPA. In that case, the court concluded that EPA did not need to establish minimum standards for MS4s and requiring strict compliance with water quality standards would be a minimum standard requirement. The court held that not requiring compliance with section 301(b)(1)(C) was consistent with its earlier ruling and supportive of its reading of section 402(p) as not requiring controls beyond the vague MEP standard.

The Ninth Circuit further determined that although Congress did not require strict compliance with section 301, EPA or an authorized state could, in its discretion, require such compliance. Section 402(p)(3)(B)(iii) also states that the permit "shall require... such other provisions as the Administrator or the State determines appropriate for the control of such pollutants." The permitting authority may determine that strict compliance with water quality standards is necessary, or that less than strict compliance is appropriate. This holding is important because leaving discretion to require more stringent limitations ensures that EPA maintains a place at the table with respect to whether MS4s must comply with water quality standards.

Harmonizing Sections 402(p)(3) and 301 : It is not entirely clear what Congress intended in section 402(p)(3). The juxtaposition of the industrial standard, which includes a reference to section 301, and the MS4 standard, which does not mention section 301, suggests the interpretation that section 301 does not apply at all to MS4s. This is the interpretation the Ninth Circuit accepted. However, this is only the Ninth Circuit's interpretation; this is still a live issue in the other circuits, and there is support for EPA's original position that MS4s must comply with water quality standards as required by section 301(b)(1)(C).

First, when interpreting a statute the court must observe a "fundamental canon of statutory construction" that a court is not to confine itself to the particular provision at issue, but rather "the words of a statute must be read in their context and with a view to their place in the overall statutory scheme." This requires that a court "interpret the statute `as a symmetrical and coherent regulatory scheme,' and `fit, if possible all parts into a harmonious whole.'" Read together, the text of sections 301, 401, and 402 suggests that MS4s must comply with water quality standards. Section 402(a)(1), which applies to all NPDES permits, requires that permits be issued in accordance with applicable provisions of section 301, which includes technology-based standards, as well as the mandate that permits shall include more stringent limitations necessary to meet water quality standards.

Section 401 contains the requirement that federally issued permits must be certified by the state to comply with state water quality standards, allowing the state to condition issuance of the permit upon imposition of additional requirements (beyond technology-based standards) to ensure compliance. These two sections demonstrate a CWA policy that NPDES permits are to ensure attainment of water quality standards. This is the CWA "safety-net"; if the technology based standards cannot achieve water quality standards, then more stringent limitations shall be required. In section 402(p)(3), Congress clearly drew a distinction between industrial discharges, which must comply with both the technology-based and water quality-based requirements of section 301, and MS4s, which are required to meet MEP.

However, it does not necessarily follow that Congress intended to waive all of section 301's requirements as to MS4 NPDES permits. It is possible that Congress intended for MEP to modify only the technology-based standards in section 301, but that MS4 permits must still contain requirements necessary to achieve water quality standards. This second reading is more in concert with the general rule that NPDES permits must ensure compliance with water quality standards and with the overall CWA policy of attaining water quality standards.

Further, the interpretation adopted by the Ninth Circuit assumes that section 402(p)(3) impliedly waived section 301(b)(1)(C)'s requirement for MS4s because as a general rule, all NPDES permits are subject to the requirements of section 301 (b)(1)(C). However, it is a rule of statutory construction that implied repeals of statutory sections by later enacted provisions are disfavored, unless the two provisions are in "irreconcilable conflict" or the later enacted section "covers the whole subject of the earlier [section] and is clearly intended as a substitute." In either case, "the intention of the legislature to repeal must be clear and manifest."

Here, the language of section 402(p)(3) is not in irreconcilable conflict with section 301(b)(1)(C). An MS4 could comply with both provisions, as section 301(b)(1)(C) only serves as a supplement to the MEP standard when those controls are not sufficient to attain water quality standards. Also, section 402(p)(3) is not a clear substitute because it does not address the issue of whether MS4s must comply with water quality standards. The interpretation that section 301(b)(1)(C) requirements apply to MS4s harmonizes the two provisions and gives effect to the overall policy that NPDES permits are to provide a means for attainment of water quality standards.

This reading is also more in concert with the general focus of the Water Quality Act of 1987, which was to improve compliance with water quality-based requirements of the CWA. This focus is demonstrated by the addition of several sections to the CWA, including section 303(c)(2)(B), which requires states to adopt criteria for toxic pollutants that may interfere with designated uses; section 304(l), which requires states to list waters in nonattainment with water quality standards, to identify the causes of nonattainment, and develop control strategies to bring those waters into attainment; section 319, which addresses controlling nonpoint pollution that is contributing to nonattainment of water quality standards; and section 402(o), which is an anti-backsliding provision preventing relaxation of effluent limits in later NPDES permits, including limits imposed under section 301(b)(1)(C).

The storm water provisions were a large part of the Water Quality Act amendments and it seems a strange result that those provisions would contain an exemption to the general rule requiring NPDES permits to ensure attainment of water quality standards without Congress explicitly acknowledging that result. Although the Ninth Circuit did not find this reading persuasive, the court's reasoning is open to question because the meaning of section 402(p)(3) could easily be interpreted as ambiguous because both interpretations are textually supported by the statute. If the court had determined that the section was ambiguous, it would have proceeded to step two of the Chevron analysis and would likely have found EPA's interpretation reasonable—an interpretation that required MS4s to comply with section 301(b)(1)(C). And under EPA's regulations, NPDES permits must ensure attainment of water quality standards, a result that EPA's interim policy does not appear to guarantee.

The troubling aspect of the Ninth Circuit's ruling is that EPA or a state permitting authority could decide to abandon entirely the policy of MS4 discharges attaining water quality standards. Because EPA has only provided vague and highly discretionary guidance as to what satisfies the MS4 technology standard of MEP, and because compliance with an MS4 permit is based solely upon good faith implementation of a SWMP, there is no statutory or regulatory protection of the goal of attaining acceptable water quality in MS4 discharges.

Status of the MS4 Programme

Of the 1059 MS4s subject to regulation under Phase I, 1017 have submitted permit applications and nine are not participating. As of December 2000, federal district courts have decided only two compliance actions. The majority of Phase I MS4s are complying with the application requirements to develop a SWMP designed to control the adverse impacts of storm water discharges. However, the SWMPs' descriptions of BMPs vary in detail and specificity, so that actual improvements in water quality may vary greatly among MS4s. Also, EPA has not conducted a nationwide review of SWMPs or measured the effectiveness of implemented BMPs on a large scale, resulting in limited information on whether implementation of BMPs is enabling MS4s to attain compliance with state water quality standards. Much of what EPA has learned about BMP effectiveness has been through the research and studies of outside organizations, which are based on case studies, limited monitoring research, and anecdotal evidence.

Storm Water Best Management Practices (BMPs)

A storm water BMP is a cost-effective strategy, measure, or engineered structural control designed to control the quantity and improve the quality of storm water. Each type of BMP has limitations on its individual ability to improve water quality, so a variety of BMPs must be implemented based upon the conditions of the locality. These conditions may include length of dry periods, average rainfall intensity, soil types, land use patterns, or the amount of impervious surface area. Storm water BMPs fall into two main categories: nonstructural and structural.

Nonstructural BMPs are "institutional, education[al], or pollution prevention practices designed to limit the generation of storm water runoff or reduce the amounts of pollutants contained in the runoff." These BMPs include a variety of practices that can reduce storm water pollution. For example, changing traditional land use patterns through growth management and conserving open space within developments reduces the overall amount of impervious surface, thus preserving natural processes that filter out pollutants and recharge groundwater. Public education and outreach operates to prevent pollution by changing harmful individual and commercial practices, such as improper pollutant disposal and pesticide and fertilizer overuse. Industrial, commercial, and municipal good housekeeping measures are used to prevent pollutants from being washed into storm drains by limiting or preventing exposure to rain, storm water, or wash water. An important and required nonstructural BMP is illicit discharge detection and elimination, which can greatly reduce some of the worst storm water pollution.

Structural BMPs are site-specific engineered and constructed treatment systems that improve the quality, or control the quantity, of storm water at the point of generation or the point of discharge. These BMPs have limited site-specific benefits that are most effective in highly impervious areas and where prevention practices alone are not enough to improve storm water quality. Typically, structural BMPs fall into four categories: detention practices, biofiltration practices, infiltration practices, and filtration practices. Detention systems collect and temporarily store storm water and then discharge it through an outlet into a water body. Biofiltration and bioretention measures use plants to filter the storm water, slow the water down so contaminants can settle out, and then allow the water to infiltrate into the ground. Infiltration systems temporarily store storm water in basins, so that the water can slowly infiltrate into the ground below. Filtration measures use media such as peat or gravel to filter out pollutants from the storm water.

BMPs are most effective if a proper variety of controls are implemented as part of a comprehensive plan. The benefits of individual BMPs are site-specific and depend upon a number of factors, including the number, intensity, and duration of wet weather flows; pollutant removal efficiency; the water quality and physical condition of the receiving waters; and the existing and potential uses of the receiving waters. There are limited studies on the overall effectiveness of BMPs in achieving water quality standards, but a proper selection of BMPs has the potential to reduce pollutant loads in an MS4's discharges.

BMP Effectiveness

BMP efficiency is the ability of a BMP to remove pollutants from runoff; effectiveness is the resulting actual improvement in water quality or aquatic habitat. Most studies on BMPs have focused on their pollutant removal efficiency, which is difficult to translate into actual effectiveness in terms of improving water quality. However, characterization of local water quality and the implemented BMPs' effect on that water quality is key to determining the success of an MS4's SWMP.

Nonstructural BMPs : Nonstructural BMPs are cost-effective pollution prevention measures that are central to a successful storm water program. These measures can have positive effects on water quality, but those effects are not being successfully monitored in a way that translates into quantitative showings of improved water quality. It is difficult to reduce the success of nonstructural BMPs to quantitative measures, such as pollution reduction and improvements in receiving stream water quality, because nonstructural BMPs focus on pollution prevention, changing land use practices, and individual habits. A few case study municipalities have attempted to quantify the improvements that result from implementation of these measures.

There is little quantitative data on the effectiveness of education and outreach. Measures that have been used in an attempt to quantify the effectiveness of these types of programs include the number of flyers distributed over a given period, the number of radio or television broadcasts, the number of public workshops held per year, the percentage of storm drains stenciled, and the number of volunteer monitoring and stewardship groups formed.

These measures do not reflect actual pollutant reductions in storm water, but these outreach efforts are important to reduce pollution from harmful individual practices, and a successful storm water program depends upon public support and participation. Outreach efforts must be regularly evaluated to ensure that people are receiving the education and changing their behaviour accordingly; these follow up evaluations may be the best measure for effectiveness. Some MS4s have made attempts to correlate education programs to the reduction of pollutants in a water body—for example, Minneapolis, Minnesota estimated that concentrations of pesticides in Lake Harriet dropped between fifty-nine and eighty-six percent due to outreach efforts.

Recycling and source controls are also important pollution prevention measures and may be evaluated by the quantity of materials recycled. Sacramento, California treats thousands of gallons of wash water each year in an effort to reduce the introduction of liquid wastes into its MS4. Fort Worth, Texas annually destroys or recycles 50,000 gallons of household-generated toxic liquid waste that might otherwise have been dumped into storm sewers. However, water quality improvements often cannot be extrapolated from these data, nor do many MS4s attempt to monitor for water quality improvements resulting from these programs.

The section 402(p) requirement that SWMPs provide for detection and elimination of illicit discharges into an MS4 has effectively reduced pollutant loads where a comprehensive program is in place. The most effective illicit discharge elimination programs combine education and enforcement efforts. Education informs businesses and individuals about the need to prevent illicit discharges; enforcement is the motivation to pay attention.

In Dover, New Hampshire the repair of a single illicit cross-connection improved the water quality from a storm sewer outfall, resulting in shellfish bed improvement of over ninety-nine percent based on measured enterococci bacteria. Boston, Massachusetts reduced its MS4 pollutant discharge and improved dry-weather water quality through its storm sewer inspection and dry-weather monitoring program. In Portland, Oregon, regular monitoring prevents wastes from reaching the Willamette River. Portland estimated that in 1996 the program prevented the discharge of 250 pounds of total suspended solids and biochemical oxygen-demanding material, 40 pounds of total nitrogen, and 10 pounds of phosphorus.

These limited case studies illustrate that the use of nonstructural controls can have a positive impact on the quality of MS4 discharges. These types of controls share the benefits of being cost-effective and involving the community through stewardship groups and education, which are important to raise awareness about the potential adverse impacts of everyday activities, such as car washing and fertilizer use. However, these methods have serious shortcomings. Education and recycling programs require regular evaluation to ensure they are reaching the targeted audience and changing their behaviour. Many outreach efforts have been poorly conceived and unsuccessful, making them unreliable as pollution prevention measures. Absent water quality monitoring, the effectiveness of nonstructural controls is speculative and cannot be relied on as effective measures to reduce pollutant loads. In addition, nonstructural methods alone cannot achieve attainment of water quality standards, particularly in communities with high percentages of impervious areas.

Structural BMPs : The use of structural BMPs can effectively remove high quantities of particular pollutants, but actual load reduction varies according to the type of pollutant measured. EPA found it was common to report between a twenty and eighty percent pollutant reduction depending upon the control used and the pollutant measured.

Structural BMPs remove pollutants in two ways: through natural processes that degrade or recycle pollutants and through physical removal of pollutants with BMP maintenance. Their effectiveness can be determined by reductions in pollutant loads and storm water flow rates and volume. However, most evaluation methods for structural BMPs focus on pollutant removal efficiencies, rather than effectiveness. Monitoring ambient water quality or bioassessment of aquatic habitat is necessary to accurately ascertain the effectiveness of a SWMP.

NRDC's case studies provide examples of the potential benefits that structural controls can provide for MS4s. In Austin, Texas, the installation of storm water detention ponds in mixed-use developments reduced annual sediment discharge by several tons and nutrients forty-four to sixty-five percent, compared to unmanaged conditions. The ponds also improved oxygenation and provided flood and erosion control through the slow release of captured runoff.

In Montgomery County, Maryland, structural BMPs prevented twenty-three percent of sediment loadings and twenty-seven percent of nitrogen loadings from entering streams in 1998. Prince George's County, Maryland reduced concentrations through structural BMPs of nutrients by twenty to eighty percent and concentrations of metals by forty to ninety-nine percent. In a Florida study of twenty-two wet ponds, the effluent discharge from the ponds complied with water quality standards at least sixty-five percent of the time.

Structural controls, although often costly, may be the key to achieving acceptable levels of pollution reduction for most urban communities. Case studies illustrate that the appropriate implementation of these controls results in water quality benefits. However, the implementation of any BMP on formerly uncontrolled storm water discharges is likely to result in water quality improvements. Nevertheless, several weaknesses of the storm water program must be addressed to ensure continuing water quality improvement.

Weaknesses of the MS4 Program

In its Phase I report, EPA found that successful storm water programs share three common traits: public support for storm water programs, planning to prevent future storm water problems through zoning and development requirements, and the identification and targeting of sources of pollutants of concern (POCs) for the local water body. However, even SWMPs containing these "successful" attributes may not ensure attainment of water quality standards because of major weaknesses inherent in the MS4 program.

Monitoring for Effectiveness

Monitoring efforts in the MS4 program fail to identify whether storm water controls are resulting in the attainment of water quality standards, either locally or nationally. Appropriate monitoring confirms the connection between storm water discharges and POCs that are impairing water quality. Sampling and characterization of dry weather discharges often will indicate the presence of POCs—an important step for illicit discharge detection and elimination, which has been effective in protecting storm water quality. Characterization of wet weather discharges and ambient water quality are important steps in determining the effectiveness of implemented BMPs. Further, the lack of comprehensive water quality data with respect to storm water limits the ability of agencies or citizens to enforce compliance.

This process of characterizing the general ambient water quality and identifying POCs and their sources is costly and time consuming. Municipalities have complained that monitoring results in large expenditures without resulting in commensurate benefits. EPA has estimated that "a minimum of ten years of monitoring both ambient and storm water discharges is necessary for each MS4 to be characterized (the time required to account for natural variation and to characterize other pollutant sources)."

Also, there is no standardization of monitoring and evaluation techniques, and the data gathered so far does not adequately characterize the impacts of storm water on ambient water quality. Standardized monitoring must be implemented to assess the effects of a particular MS4 on receiving water quality. In some circumstances, the best factors for determining the success of a SWMP may be indicators such as receiving stream morphology, habitat, and biological communities, rather than measuring gross pollutant loads.

Despite these difficulties, substantial steps are needed in the characterization of storm water quality so that appropriate BMPs are implemented that address the needs and conditions of a particular watershed. If MS4s are going to achieve compliance with water quality standards—which is an overall goal of the CWA—the focus of storm water and BMP monitoring needs to shift from efficiencies to effectiveness. Regardless of the number of nonstructural or structural controls an MS4 implements, if the selection and implementation of those controls are not effective at improving ambient water quality, the MS4 will likely remain in noncompliance with water quality standards.

Measurable Goals and Performance Standards

The lack of set minimum performance standards for MS4s may be hindering the attainment of water quality standards and the determination of whether the MS4 program is successful. Although flexibility in the storm water program is valuable and necessary, without definitive baseline standards it will be difficult to enforce compliance with individual permits, and the success of SWMPs will remain difficult to measure.

NRDC concluded in its storm water study that measurable goals for MS4s are important because "they provide a standard by which citizens and the municipality itself can judge municipal efforts. Clear standards are essential to maintaining accountability." However, under the current regulations, "[t]he plans set forth in permit applications are just that—plans." Without enforceable standards, effective implementation of those plans cannot be guaranteed.

Additionally, the Ninth Circuit decision in Defenders of Wildlife v. Browner not only confirms EPA's decision not to require achievement of measurable standards, but allows CWA permitting authorities to refrain from requiring compliance with water quality standards at all.

This result suggests that the extra effort and expenditures necessary to measure ambient water quality may provide the necessary disincentive for states not to require water quality monitoring, given that water quality standard compliance is not mandated. This lack of substantive guidance and measurable goals may result in MS4 permits that do not seek to improve upon the early demonstrated benefits of first round BMPs, despite the fact that ambient water quality standards may not be achieved.

Watershed Management and TMDLs

The current trend in resource management is a movement towards watershed wide integration of land, air, and water controls. Watershed management focuses upon the health of the entire watershed and then tailors resource management programs to achieve those health-related goals. EPA has indicated a need for better watershed management and integration of wet weather controls on a watershed basis.

Reflecting this trend, Congress recently amended the CWA, granting EPA the authority to provide technical assistance and grants to treatment works initiating wet weather watershed pilot programs. A successful storm water program, particularly one that takes into account the realities of watershed management, requires participation in planning and decision making by all affected communities, particularly because nearby unregulated MS4s affect regulated MS4s' ability to attain water quality standards. Integrated control of all pollution sources within a watershed or sub-watershed would be beneficial to achievement of water quality standards. This effort requires cooperation across community and utility boundaries.

Use of TMDLs for an integrated management of water quality-limited waters is an attractive possibility. TMDLs look broadly at the cumulative effects of discharges on a watershed, taking into account all pollution sources, including nonpoint sources, when assessing the amount of pollutants a waterbody can absorb. Integrating TMDLs into watershed management is growing in importance. The TMDL provision of the CWA was largely ignored by the states until of series of citizen suits, which compelled EPA to create TMDLs for water quality-limited waters if the states fail to do so. EPA believes that TMDLs will have a significant role in converting conventional pollution control to watershed management.

For an effective watershed approach, all CWA requirements for a particular area should be combined into a unified effort, which includes integrating TMDLs into MS4 programs. This does not mean that the MS4 permits would have to contain numeric effluent limits. TMDLs may also specify "habitat parameters," such as indicating high erosion areas or lack of riparian vegetation, which are then linked to specified management measures. These TMDL measures could provide enforceable, measurable goals for MS4s located on water quality-limited waters and would facilitate a watershed management approach. This approach would also require the type of monitoring efforts now lacking in the MS4 program—assessing ambient water quality and characterizing storm water discharges impact on water quality and habitat.

Regulation of MS4s under Phase I of the CWA storm water program is reducing pollutant loads in storm water discharges. Appropriately implemented and maintained, BMPs can accomplish attainment of water quality standards. The difficulty lies in whether the manner in which the storm water program is being implemented is meaningful. The stated goal of the program is to achieve water quality standards, but without mandated performance standards and storm water quality characterization, it is uncertain whether this goal can be achieved.

The Ninth Circuit not only has upheld EPA's policy not to require any minimum performance standards or measurable goals, but it has placed the decision to require more stringent limitations necessary to comply with water quality standards entirely within the discretion of permitting authorities.

This discretion has been exercised only minimally by EPA in its interim permitting policy, which does not require more stringent limitations, but rather endorses the use of "better-tailored" BMPs in subsequent permits where "appropriate." This type of permitting is highly discretionary with each permit and does not guarantee that BMPs effectively addressing water quality problems or POCs will be implemented.

The history of the CWA is littered with instances of state and EPA nonenforcement of state water quality standards, resulting in citizens filing lawsuits to compel compliance. However, the lack of minimum measurable standards—technology or water-quality based—in the MS4 storm water program and the lack of comprehensive water quality data may potentially foreclose this avenue of ensuring CWA compliance.

The first step in creating a storm water program that ensures meaningful implementation of pollution controls is to address the information gaps in water quality and monitoring. Both municipalities and EPA have recognized that the current system does not enable entities to determine the effectiveness of implemented BMPs. EPA and the American Society of Civil Engineers (ASCE) are currently addressing this information gap. In a joint effort, they recently developed a standardized monitoring system for the evaluation of BMPs to be included in the BMP database.

However, this effort still reflects a focus on site-specific measures, rather than on overall water quality assessment for the MS4. Once a more meaningful monitoring system is established, this will suggest the direction the storm water program should take. Ultimately, MS4 permits should be tied into TMDLs that are developed for water quality-limited waters. This would also provide measurable standards by which compliance with the CWA could be evaluated and would facilitate a watershed regulatory approach. The MS4 storm water program has the potential to improve the quality of many polluted water bodies, but without minimum standards and required compliance with water quality standards, that potential may not be realized.

5

Complex Attachments to Water-places

This chapter explores water as central to defining complex attachments to place, an approach that is concerned to contribute to discussion beyond water's conceptualisation as a vital and natural resource in need of environmental management. The value of an active discourse and practice that emphasises water as a resource to be managed is obvious, for example, in cases where a sudden increase in water flows has the potential to transform rivers into floods with serious consequences for humans and other species. My interest, however, is to consider whether discussion of attachment to water sources in the Kimberley region of northern Western Australia, can show how, and to what extent, contemplation of water/human relationships might fruitfully facilitate understandings of 'place-based cultures' (Escobar 2001:142).

While notions of 'attachment' are sometimes conflated or confused with concepts of 'belonging'—about how and why persons (and/or native and introduced species, and so on), reveal or describe a sense of belonging to certain places and environments—the emphasis here is primarily on attachment. Like my co-contributors (some of whom examine concepts of belonging), I am concerned to extrapolate varied interpretations about a key environmental trope—in this case, water—to foster a deepened understanding not only about a precious environmental resource, but also about social life, cultural politics and material struggles in an Australian setting.

Concentrating on the remote Kimberley region's Fitzroy River, and examples where water sources sit on their own or become agents of transformation (for example, as rain or in food preparation), I begin with a consideration of guiding themes and then turn to Australian literature about people's relationships to water, especially to texts concerned with indigenous perspectives. A selection of Kimberley ethnographic data, a region where I have undertaken fieldwork since 1981, follows. Discussion of this material is used as a backdrop to canvass the ideas of Escobar (2001) on attachments to place, and of Tsing (2002, 2005, 2007) on the creation and outcome of 'friction' when cultural and environmental perceptions and activities collide. I conclude with the claim that studies of human/water relationships benefit when they are examined within the context of a place-based-culture analysis. Such a context is characterised by a mix of attachments that happen to be about places that involve water.

Friction, Attachment, Place

Conflicts and disputes over scarce or abundant water sources are major points of inquiry in current water studies. Tsing's use of 'friction', however, has the potential to provide a more comprehensive and nuanced way to explore the complexities of disjuncture and/or resonance that occurs when proposed developments threaten fragile ecologies and the life styles of nearby populations. This is especially the case on projects involving indigenous groups. Tsing (2002:472) encourages intellectual and ethical space to help make visible, rather than leave in unexamined form, problematic ideologies about environmental and indigenous interests. Tsing adopts the expression 'friction' to encompass elements of what she describes as the 'grip' and 'irritation' that interact to, on the one hand, facilitate and, on the other, limit, productive engagement and positive outcomes when environmental matters are a point of departure (Tsing 2005, 2007:58).

Tsing's (2005, 2007) work is mostly focused on Indonesia, but her conceptualisation of 'friction' facilitates a fertile framework from which to examine issues outside south-east Asian locations. Friction also sits nicely alongside 'traction'. Whilst arguably a term that emphasises the value of 'grip', traction is increasingly used in Kimberley settings to indicate a desire for concrete commitment and action. The comments of a local indigenous resource agency coordinator (identified here as 'Leo') give voice to my claim:

What people [referring to indigenous communities] around here need is a bit of traction—government and industry getting a grip on what people are talking about, what their needs are. Consultation means nothing around here without grip, without proper traction, without something or someone actually taking hold (Fieldnotes, Kimberley, November 2007).

Leo's use of traction is not too distant from Tsing's 'friction', albeit in contrasting contexts. His comments, made to myself and two indigenous people present, took place after a phone conversation with a major developer who had offered to put a large amount of funding into local initiatives—as long as any plans that were put forward accorded with the developer's time-line. In later discussion, it became evident that both the phone call and the caveat were not unusual occurrences and, despite a need for funding, the offer had implications attached to it that were not straightforward, a point that indicates how Tsing's use of 'friction' might be understood.

This is most apparent when Tsing argues that evidence of 'grip' and 'irritation' as 'friction' emerges when examining social and environmental movements and their relationship to the corporate sector. In the above case, for instance, it is obvious that whilst the indigenous organization was offered some sort of 'grip' in the form of financial support, the offer was not without the 'irritation' of an unrealistically short timeframe. As Tsing notes, however, while there is little evidence to show that social, environmental or indigenous movements have the ability to 'displace the hegemony of private property and capitalist development', a certain 'messiness' (2002:472, 476) exists in such situations, such that 'not all indigenous peoples support environmental causes; some are in active conflict with conservation' (Tsing 2007:55). Escobar (2001) seems less concerned with the problem of inflexible categories and the characterization of issues that lend themselves to an analysis based on friction.

His concern is on practical engagements with, and resistances to, the unwanted development of places people hold dear for a variety of culturally endowed meanings. Escobar (2001: 147) is critical of studies that over-emphasise mobility and de-territiorialization to the extent that 'place-based practices and modes of consciousness for the production of culture' are overlooked. A proponent of the need to problematise unqualified generalisations, he also aims to de-centre the uncritical use of 'Western' forms of knowledge. Concentrating on the work of theorists such as Auge (1995), whose binary construction of 'non-place' tends to undermine concepts of place as multi-sited, contingent and relational, Basso (1996) and Bender (1993), among others, Escobar claims that contestations about place help to unravel the tensions embedded in material struggles. Such a process inevitably attracts consideration of local cultural beliefs and practices (2001:170171). Adding a further layer of meaning with regard to how concepts and constructs of place are constituted, Escobar cogently claims that local knowledge can also be read as political knowledge: 'Local knowledge is a place-specific way of giving meaning to the world'.

Whilst commending the literature on place as a site of sensory and phenomenological attachments, Escobar's main concern is with the politics of place-based cultures. He shows how attachments to place are constantly created and re-created by people's social and political engagement with that place, such as via active resistance to protect an area against development projects predicted to cause harm. In Escobar's parlance, social, cultural and political attachments are formed and become meaningful through active dialectical processes of 'connectivity, interactivity and positionality'. Such an approach logically includes consideration of divergent identities that emerge within 'the dynamics of place, networks and power'.

Giving prominence to the need for the cross-fertilization of political, economic and cultural approaches within the theories and practices of a political ecology framework, Escobar (2001:153) opines that an inter-mix of attachments to 'place' is given meaning by diverse groups of women and men through active political involvement. Escobar is also concerned to analyse attachments to place as an outcome of beliefs and practices where interactions between persons and their environments are in the constant process of mutual transformation. Such a process accords with Tsing's view that social (including indigenous or 'subaltern' to use one of Escobar's terms) and environmental movements fluctuate over time, a process that results in them being difficult to classify coherently, especially in circumstances where the local and global interconnect.

Read separately or in tandem, what these authors provide is a creative and dynamic framework to explore how and when description of water/human relationships might benefit from an approach that privileges culture as place-based attachment. That social, indigenous and environmental movements and issues can be constructively interpreted through the intertwined notions of friction and traction, is also a point of inquiry.

Next I outline a selection of Australian literature in which it is plain that contemplation of water prompts a rich and diverse range of studies. Most of these texts concentrate on Aboriginal and Islander groups and inter that water-places are significant religious and economic sites, sources of survival, and places endowed with cultural and political meanings.

Human/water Relationships

A visible social science and humanities research and publishing interest in water/human relationships has become increasingly evident in the past decade. Such a key epistemological shift has the potential to ensure that cultural description and analysis, and past and present human activities, is regularly embedded in studies of the environment. It remains the case that more independent anthropological and cross-disciplinary research needs to occur to augment environmental, biophysical and natural science inquiry, but there is no doubt that a substantial volume of more explicit culturally attuned work with water at its centre is emerging. As well as contributions in this volume, monographs in the field of Australian, Pacific and south-east Asian water studies include those by Balint (2005), Brearley (2005), Leybourne and Gaynor (2006), and LahiriDutt (2006). This list does not include works outside these geographic areas, nor does it acknowledge texts focused on relational inquiries, such as that by Beresford et al. (2001) who discuss the salinity crisis in Western Australia from social science perspectives.

Recent trends are especially evident in research being conducted with and/or about Aboriginal and Islander Australians, in part because of the High Court's 1992 Mabo Decision where the Meriam Islanders in the Torres Strait off the northern Queensland coast were found to have long-term cultural affiliations with not only land but also with fresh and saltwater sources on the Island of Mer. Altman (2004), with regard to water issues, canvasses property rights and interests in native title provisions and land-use agreements, and Bagshaw outlines the Kimberley Karajarri native title claim where water featured prominently.

Notwithstanding the water-based research that is now occurring in native title claims and cross-disciplinary environmental research more broadly, a selection of material involving Australian indigenous groups reveals a spectrum of relational topics ranging from past and present subsistence and customary law marine activity, to spiritual affiliations, beliefs and rituals, to environmental implications of water planning policy. Of particular interest is that each writer indicates, without always making it explicit, water-inspired beliefs and practices as these accord with the making and meaning of place-based cultures. Bayly (1999), for instance, brings anthropological, archaeological and historical data together to show the ability of indigenous groups in the Western Desert bloc to 'read' the landscape and find surface and groundwater in desert locations, a quality that disadvantaged non-indigenous travellers and explorers in the distant and recent past whose cultural experience did not involve such engagements.

Like other commentators and researchers, Bayly argues that, historically, indigenous migration was largely dictated by knowledge about the presence or absence of water, and water's ability to transform the environment. Rose (2004) is also concerned with how water sources were and are central to sustaining life and land, especially in the desert during prolonged drought. Rose, in a way comparable to Escobar's (2001) usage, describes this process as one of 'connectivity' that includes ground water, surface water and rain mediated through the Rainbow Serpent or mythical snake deemed responsible for water's creation. As Rose makes plain, indigenous groups had in the past and continue in the present to have a responsibility to care for all water sources. Any negative interventions (such as the uncontrolled drinking of water by cattle, or the implications of mining, both of which can result in contamination) pose major problems for people (indigenous and non-indigenous or 'settler') and the culture-based landscape in which they are embedded.

Writing specifically about the Gulf of Carpentaria in northern Queensland, Trigger's (1985) research takes a more fine-grained ethnographic approach to explore the naming of fresh and saltwater sources. He argues that water is a highly significant feature of the environment: water acts as a 'key signifier of meanings attributed to nature that extend beyond water places themselves to encompass the landscape more broadly'. Langton (2006), with supporting cultural data gathered from groups with traditional affiliations to the Laura Basin area of Queensland's Cape York Peninsula, claims that people relate to water through the regular and interconnected cycle of 'earth, wind, fire and water'.

Morphy and Morphy (2006), sharing some resonance with Langton's discussion of the relationship between spirituality and the environment, describe how successive generations of Yolngu at Blue Mud Bay in the Northern Territory's Arnhem Land enact rituals to enhance contrasting spiritual connections with salt and freshwater places. In a framework conceptualised as 'eco-mythology', where the authors are keen to make a distinction between interactions with land and with water, Morphy and Morphy discern that an imposed dualism between salt and freshwater 'fails to capture the complexity of the environmental, ecological and mythological relationship between the two' (2006:74). Morphy and Morphy also observe that 'The Yolngu have an overall model of water as process', a claim that is explored through the use of metaphor to explain fluctuating patterns of water ownership, ritual performance, local interactions and attachments to water-based places. Magowan (2007) writes too about Yolngu attachments to water in a way that highlights the metaphorical re-cycling of practical and symbolic interconnections. These are evident in description of water currents, long-term ecological knowledge, music production and performance, and an all-embracing spirituality derived from the body of religion and law known as the Dreaming (2007:133-141).

Whilst I am less concerned in this chapter with people's relationship to saltwater sources and marine tenure than with freshwater rivers and related waterways, Bradley's (2006) research is useful here because it attends to how (salt)water influences identity formations and kinship interactions among the Yanyuwa living in Queensland coastal communities. In such a process, attachments to (salt)water-places are privileged. His work cogently embraces the sociality and interdependence of humans with their environment, a summation that specifies cultural attachments to saltwater places that have been reproduced throughout time.

According to Bradley, identity is constructed around named and relational saltwater sources and interdependent sea-living species. Similarly to other authors whose work, Bradley presents rich ethnographic data to consider water/human issues as these explain Yanyuwa social and cultural life. He argues that the Yanyuwa see no distinction between land and water and that 'the sea and seagrass beds... [are all part of]... geographic land units' (Bradley 2006:128). Indicating the importance of saltwater interactions, alongside analyses based on the qualities of 'sea knowledge' and attendant power relations, Bradley concludes that Yanyuwa identity is largely founded on 'enduring emotional links between people, sea-country and many of the creatures which inhabit the area'.

Goodall (2002) contemplates issues of water and identity too, and in a way that accords with Escobar's (2001) place-based cultures framework. Her research also facilitates a return to freshwater rivers and associated waterways, in particular how forms of attachment were heightened when water resources were threatened. Focusing on indigenous and non-indigenous (or 'settler') groups in New South Wales who contested irrigation for cotton production on the Darling floodplain, Goodall shows how the project impacted negatively on local groups and the broader social and political life of nearby towns. Cognisant of the intertwined relationship between land and water, she observes that people construct attachments to place because of their (often long-term) lived experience with it, their social history and livelihood, and a series of remembered, sentimental associations. In such places, water is a key social as well as environmental element because 'it is water which creates and replenishes the land and the soil' (2002:36), a circumstance that influences people's work practices and social activities at nearby lands and waters.

A further dimension to Goodall's research is that demographic changes are shown to have occurred as a result of introduced irrigation and cotton production. Demographic change has obvious implications for social change. In the case presented by Goodall, one of these was that many senior citizens moved away from the area once the development was established. Their relocation resulted in the loss of long-term local knowledge, support and experience (2002:47-48). Goodall also shows how people's hitherto positive attachments to places associated with the floodplain were frequently replaced by 'a set of fears so deep they can only be expressed in terms of overturning the most basic rule of nature'. The struggle over a place around which water was central to how people constructed and maintained their identity, enacted social relations, undertook past and present work-practices and sustained sentimental associations, resonates with Escobar's ideas on how and when attachments to place are created and re-created. Evident also in Goodall's material is that indigenous groups and non-indigenous graziers were united in their opposition to the project, and that friction existed between them and the developers. This situation also reveals that attachments to place were amplified when the Darling floodplain was threatened; whilst the floodplain mattered, it was the struggle over it that mattered most.

Among the issues that emerge when evaluating water/human relationships in Australia, is that most studies highlight the intrinsic, local nature of the land/water/people interaction and connection. It also becomes clear that salt and freshwater sources should not be treated as one and the same. That a hydro-and ecological change in how a water source is used can lead to a change in how people relate to it encompasses a range of culturally complex issues, including that water is engendered with a variety of meanings. Identity formation and kinship affiliation can also be determined through research on water, as can knowledge about contested usage, and patterns of migration to and from temporary and permanent water places. Each study implies, and sometimes explicates, ways in which local groups become attached to sources of water beyond water's obvious nourishing, life-giving force. This is especially the case when water sources are endangered, and cultural ideas, beliefs and activities collide.

The Kimberley's Fitzroy River and associated waterways present excellent scope to evaluate matters related to how and when attachments to place are formed in the way Escobar describes. It also presents data to explore Tsing's conceptualisation of 'friction'. I begin by explaining a little of the Kimberley, in particular the Fitzroy Valley in the West Kimberley landscape, and then turn to a case-study which features the Fitzroy River.

Kimberley Water-places

The Kimberley is a vast and rugged region of over 100,000 square kilometres with temperatures ranging between 20 and 45 degrees Celsius and landscapes alternating between desert, ranges and coastline (Brown 2004; Storey et al. 2001). A place of many contrasts, the Fitzroy River and its tributaries (including the intersecting Margaret and Leopold Rivers) is central to understanding the West Kimberley landscape from different environmental, geographic, economic and cultural vantage points. One of the last unregulated rivers in Australia, the Fitzroy River traverses from the West to the East Kimberley across 85,000 square kilometres of landscape (Storey et al. 2001).

Published and unpublished research in the field of native title discusses indigenous affiliations with a range of Kimberley water sources, such as rivers, soaks, creeks, pools, lagoons and 'living water' (known locally as jila) and so on. Research available in other material describes the manifold ways in which groups manage and make meaningful interconnections with waters and rivers. In the West Kimberley, the Ngarinyn, Karajarri, Bunuba and Gooniyandi are renowned as river owning and using groups, with members of the desert-related Walmajarri and Wangkajunga groups establishing a form of cultural co-existence in the area as the result of a 40 year process of voluntary, involuntary, and negotiated migration. A Walmajarri woman, Daisy Andrews, describes this interaction thus: 'It's really good now how we can all sit down and work together. It's really good, river and desert together, whole lot' (Mangkaja Arts, 2003:6). Other desert people evoke river-based totemic affiliations for children born post-migration, and often describe how the river people taught them how to fish. Walmajarri woman 'Ngunda' put it this way:

> We had to learn about the martuwarra [river], how to fish with string and burnt needles shaped like a little hook, we didn't have throw nets. Nothing! The Gooniyandi and Bunuba [meaning river groups] taught us how to fish—how to hunt, cook and eat fish (Fieldnotes, Kimberley, July 2006).

In parallel with other Australian indigenous groups and elsewhere, distinctions exist between salt and freshwater sources, and rivers and other forms of water (such as rain, and permanent or ephemeral waterways), continue to be conceptualised as having been inspired by the Dreaming. Home to approximately 40 Aboriginal communities and four major towns, people in the region, like the landscape of which they are an integral part, are constantly responding to change and some extremely tough social and economic conditions. Despite these conditions and, in many cases what can only be described as circumstances of poverty and neglect, indigenous rights to, and interests in, land and water continue to be based on a complex combination of patrilineal and matrilineal descent, conception sites, birth sites, totemic affiliations, historical associations and residence.

Whilst economic initiatives for indigenous groups are improving in the region, evidenced in projects such as the Manbana Aboriginal Aqua Culture Centre in Broome which is now providing study programs for local people, aquatic tours for visitors, and expanding fish markets into south-east Asia, many Kimberley groups continue to exist on government controlled social security entitlements, an economic problem with social implications that has received ongoing local attention. Opportunities that add to household incomes include contract cattle station work, intermittent involvement in tourism activities, for example as tour guides on Aboriginal land and in national parks, and the results of subsistence work.

Fishing is an important economic activity too, including the exchange of fish when one family is more successful than another after a day at the river. Education for young people also occurs via fishing with the teaching of local language names for fish during what would otherwise be regarded as recreational and economic activity. During periods of mourning when fish as a sole food source are necessary to accommodate taboos on the consumption of beef products following the death of a loved one, the catching, cooking and consumption of fish becomes vital (Toussaint et al. 2005:66).

Relationships to water also continue to be understood by most women and men as having been inspired by the Dreaming, known locally as Ngarrangkarni or Jumungkarni. Walmajarri man, Joe Brown, recounts the cyclical interconnections between the Dreaming, people, land and water:

Everything we know comes from the Dreamtime. People even in town still know how to sing story for their country. Every jila [water hole] has its own songs, story and skin groups. Without the snake underneath the water will go away. We have been looking after our waterholes and rivers for thousands of years. We have respect because we know that if you don't treat it right many things can happen. This is the reason that we need to make other people learn (quoted in La Fontaine 2006:217).

Janjin Rogers, also a Walmajarri man, adds example to Joe Brown's description. His narrative also reveals the transformative and cyclical qualities of water:

My mother was worrying about me getting thirsty. She started singing for rain, I couldn't believe that the rain would come. There was a little cloud in the sky, my mother sang it and it got really big. Then it started pouring rain on the ground, water everywhere. I was really happy. My father came back, he saw everything wet and he was happy for the good rain (Karrayili Education and Manghkaja Arts 1993:8).

Seasonal rains, whilst sometimes causing flood and community isolation for up to three weeks a year, are central to ecological restoration and to local perceptions of what is considered to be the river's natural course: annual monsoonal rains ensure the replenishment of vital resources (waters, lands, plants, animals, fish, and so on) for humans and other species. River stories also find expression in the production of art, such as paintings made for sale to locals and visitors, and for state and national art exhibitions (Toussaint et al. 2005:68-71). Paintings about water also increasingly reflect political themes, including the need to protect local rivers and waters with which people have long-term culture-based connections and attachments.

This point is elaborated below. From the insights raised here and expanded elsewhere, it can be inferred that Kimberley indigenous stories and activities regularly explicate complex matters relating to how people interact with, and subsequently form attachments to, the river and other water-places.

The water/human relationship evident for Kimberley groups also shares certain parallels with indigenous peoples in other northern Australian settings. Local groups have rights and interests in waters and lands through customary law, via forms of residence, and as a result of desert migration. The painting of water-places has added to some people's income, and the river presents a location where subsistence activity and learning occur. Less well known is how non-indigenous people relate to local water-places. A discussion of non-indigenous/water relationships is, of course, just as vital as those of an indigenous kind, especially when canvassing how frictions might emerge and place-based cultures be identified.

The Fitzroy River, Friction, and Place-based Attachments

Unlike indigenous peoples who have generations of involvement with the river's creation and protection through their own body of religion and customary law, social and political networks and economic activities, settler relationships are less well documented. In broad terms, these can be classified as having occurred more recently and incrementally through temporary and permanent residency, often involving work for government, non-government and industry (e.g. tourism, mining), and pastoral work on cattle stations. In addition to work-related interactions, activities such as fishing, camping, bird watching and recreation are regularly reported.

Such practical, economic and social activities have undoubtedly established memorable ties of affection and sociality that, on some occasions, are shared with indigenous people. Settler environmental groups are the focus group here, especially as they have been actively involved in protecting the river in particular, and the broader environment more generally, against unwanted developments. This circumstance has drawn a mix of groups and organizations together.

A timely beginning is 1996 when government and industry plans to dam the Kimberley's Dimond Gorge, and two of its major tributaries on the Margaret and Leopold Rivers, became public knowledge. With parallels to the study outlined by Goodall (2002), the aim was to establish a cotton enterprise based on irrigated water. Local organizations such as the Kimberley Land Council and the Kimberley Aboriginal Law and Culture Centre contested the development with support from national environmental agencies such as the Australian Conservation Foundation. As I have shown elsewhere (Toussaint 2008a), this situation united local groups in their opposition to the proposed developments. The regional aim, involving a diverse range of local indigenous, tourism, hospitality and pastoral groups, became to protect the Fitzroy River's waters for Kimberley people and Kimberley lands, in particular the economic value of locally-controlled tourism to the region, of which the river was a central component.

The 1996 initiative also resulted in the establishment of a Kimberley-based environmental organization that involved non-indigenous people committed to local conservation and nature issues. Called 'Environs Kimberley', the organization has become increasingly active in the region where it now has a Broome-based office, several permanent staff, and a large assortment of dedicated members and volunteers.

Whilst the 1996 plan was eventually withdrawn, in 2005 the Western Australian State Government (in concert with industry) again expressed interest in the Kimberley's extensive river system. They established an inquiry to assess the feasibility of diverting the river and/or transporting its water 3,500 kilometres south to offset water shortages in the State's southern region. Three options were investigated: a Kimberley to Perth pipeline, a canal and ocean transport. The inquiry recommended in its final report that all three options were too costly and that limited knowledge existed about the 'hydrogeology and ecology' of the West Kimberley. Despite such an outcome, indigenous and non-indigenous groups continue to be agitated that some form of river diversion is not too distant from the minds and plans of government and industry. This has resulted in a series of nuanced frictions. Before explaining this situation, it is useful to outline certain beliefs and practices that characterize non-indigenous relationships to the river, some of which parallel those of an indigenous kind.

'Sam' has lived in the Fitzroy Valley for several decades. He is one of the few non-indigenous workers regularly employed by an indigenous organization. His work often involves him in field trips to desert locations several hundred kilometres south of the town of Fitzroy Crossing. Sam's comments make plain his love of the area, especially the Fitzroy River:

I like going to the desert [hundreds of kilometres south of the Fitzroy River] for native title and heritage work. We usually travel in convoy, a few four-wheel drive vehicles travelling together. It's the jila [water holes] I'm most interested in. And then I'm happy to be back [in town] near the river. It's everything here [Fitzroy Valley]. I wouldn't be here if the river wasn't here. A lot of people wouldn't be (Fieldnotes, Kimberley, June 2006).

Sam's views can be positively compared with those of a local ecologist who worked for an indigenous resource agency and described himself as feeling 'passionate' about the Fitzroy River, a place where he regularly fished and camped on weekends. Noting the importance of the river to his work, 'Didier' commented that frictions could arise 'when indigenous people aren't actually present to explain their interests to outsiders, such as to developers...' (Fieldnotes, Kimberley, July 2006).

Despite indicating a passion for the river, Didier also stressed his long-term 'commitment' to the place and the local people, in part by talking about how 'proper, fee-paying' tourism might help to support and conserve the Kimberley's ecological and cultural environment. He commented on the importance of exploring the feasibility of commercial projects that might:

result in the environment being more sustainable, and help to get people and organizations away from a reliance on government funding. The traditional owners [meaning local indigenous people who sustained customary law rights and interests in lands and waters] would need to agree, of course, and their rights and interests in all this [lands, waters, coastlines] are not unreasonable. But we've got a long way to go, and not many resources.... (Fieldnotes, Kimberley, July 2006).

In such a positioning, Didier makes plain his concern for indigenous groups and cultural matters. But in a later conversation, he positioned himself slightly differently. Didier pointed out that whilst it was constructive to see that indigenous people were becoming increasingly involved in carefully controlled tourism projects (such as at a place called Jarlmadangah where several Walmajarri and Mangala families operated a successful tourism business), he and other ecologists were worried about the impact an increase in river cruises was having.

Run by another indigenous group with assistance from local operators and/or in conjunction with state-based conservation agencies, the boat cruises for tourists had resulted in the Fitzroy River experiencing expanding environmental flows, and the wash from these often impinged on the river's banks with consequences for wildlife. For example the nesting habits of crocodiles were affected as they had been obliged to move further inland (to less safe places) when the seasonal time came to lay their eggs (Fieldnotes, Kimberley, July 2006). Of interest in this scenario is that while Didier's comments suggested he could understand the need for indigenous groups to be involved in a tourism venture, he felt torn about such an initiative because of the venture's impact on the environment.

What was not at issue, however, was his attachment to, and connections with, the Fitzroy River as a culturally-endowed place. Whilst it is evident that non-indigenous people such as Sam and Didier value the Fitzroy River and the broader environment, this was not always the case. 'Craig', an environmental health officer with a government agency, had quite a different perspective:

I don't really care much about the environment here, like it's interesting, I guess, and I'm an environmental health officer so I get out and about a bit, into the bush, along the river system, visiting communities and stations... but it's the easy life style that gets me in. It's nothing to do with the environment, and I'm not sure how long I'll stay here (Fieldnotes, Kimberley, June 2006).

Craig's comments make clear his position. At the same time they both support and diverge from Escobar's framework about how and why attachments to place occur. At one level, Craig showed that he is attached to the environment because of the 'life-style' it provided, but at another he revealed the lack of a long-term commitment to people and place in a way that differs from Sam's and Didier's emphases. The comments of a freshwater campaigner, 'Gary', are particularly apposite here, perhaps because they help to crystallise the two perspectives. With a background in sociology and ecology, Gary's work with a Kimberley environmental agency shows how his attachments to two different rivers—one of which is the Kimberley's Fitzroy River—were reliant on a struggle to protect each resource. His personal positioning clearly shows how human/water connections were established through active political engagement:

I can see that it [the Fitzroy] is a beautiful river... but I haven't formed an attachment in the way that I did for the McArthur River [in the Northern Territory]. I was a freshwater campaign officer there too and although I only visited the river twice, I formed a real attachment to it... I felt passionately about it, partly through my experience with the local people... but also because I had to defend the river against mining and other activities that threatened to damage it. I guess you could say I had a personal experience with the McArthur... I might develop something with the Fitzroy River if I have to campaign for it. It depends a lot on what happens when I do more work on it and mix with the local people.

Coordinator of Environs Kimberley, Maria Mann, makes a similar point about how extensive involvement with a place can enhance people's social and cultural relationship with it. In a 2006 event celebrating ten years' of active campaigning, Maria was on site near a narrow Fitzroy River Gorge when asked by a journalist to talk about river campaigns: 'You have to try to imagine—if the dam had gone ahead this would all have been under water...' (quoted in Prior 2006:4). In the same article, co-founder of Environs Kimberley, Pat Lowe, said that the environmental group:

simply couldn't believe that it [damming or diversion of the Fitzroy River] would happen in the Kimberley... There's been so much land clearing and so much destruction already... Also, we were concerned about riding roughshod over indigenous peoples' rights (quoted in Prior 2006:5).

After a three day event at a Kimberley Wilderness Conservancy Camp, about 50 Environs Kimberley members were involved in activities such as 'bird watching, ecology tours, canoeing, and swimming' (Prior 2006:5). As a parting comment, Lowe pointedly claimed: 'When we were flying out here Ito the Wilderness Camp]... I felt moved to look down on the river and think we had a bit to do with saving it...' (quoted in Prior 2006:5). What these comments reveal from non-indigenous perspectives, at least, is that when a water-inspired place became a contested site, people's struggle with it, their positioning with the river as ally, resulted in a profound sense of attachment. Such an attachment cannot, of course, be confined to settler groups. As I have argued elsewhere and above, indigenous groups have for countless generations sustained a complex suite of connections to, and attachments with, the Fitzroy River and associated water-places. A perfect example to illuminate this point further can be seen in the imagery and artistry of Gooniyandi artist Butcher Cheryl who, each time the Fitzroy River was threatened, produced a series of paintings to visually position his concern. A quote from Butcher articulates his concerns as well as his cultural connections:

I have been making these things (images) for a long time. I put what I am looking at, tree, river, bush tucker, all that thing, that Ngarrangkarni (Dreaming) story too.... I know that Ngarrangkarni from nay mother's side. I know that country too from long time, like Manjoowa [permanent waterhole on the Margaret River]. There's crocodile, fish, everything there. I know this country from long time, they can't change that river country [referring to the damming or diversion of the Fitzroy River], it's no good that idea.

Whilst both indigenous and environmental groups remain concerned that efforts to dam or divert the Fitzroy River's flow have not completely disappeared, these are not the only issues they have to face. Others include the impact of large-scale gas projects, poorly managed tourism, the problem of introduced weeds, and climate change. Within this spectrum of concerns a number of frictions emerge.

For example, INPEX, an internationally owned petroleum company is hoping to secure access to locations such as the Maret Islands off the Kimberley coast to progress a major gas project, a move that has distressed conservation groups. INPEX executives have also made plain their intention to directly consult with indigenous groups in keeping with an Indigenous Land Use Agreement being negotiated in accordance with policies embedded in native title processes. Such a possibility has upset local environmentalist groups whose interest is to protect local biodiversity and the broader landscape; it also signals the kind of friction embedded in Tsing's analysis. I can see that it (the Fitzroy) is a beautiful river, but....

With different forms of water as guide, and the culturally influenced human/environment relationship as imperative, it is possible to engender ethnographic description and analysis as a means of interpreting place-based cultures. From the data provided here it is plain that Kimberley indigenous river and desert-related groups have long-term religious, socio-cultural ties to water-places, and that these are central to understanding the broader historical ecology as well as native title laws. It is also the case that the Fitzroy River, in particular, is a meaningful place for many non-indigenous women and men. The river is central to why many people live, work, stay and play in the area.

Escobar (2001) claims that places are endowed with meaning because of what occurs there, especially in situations where cultural politics are at issue. In the case discussed here, attachments indigenous groups have with the river, alongside attachments that non-indigenous groups have developed with the river over time, result in the Fitzroy River being not only an important water-place, but also a site of meaningful engagement, including struggle over its usage. Sectors of indigenous and settler societies who live in the Kimberley care deeply about the river for many different socio-cultural, economic, emotional and political reasons: the river is part of their permanent or temporary homeland, some people have customary law affiliations with the river, others fish and swim in the river for food and/or for fun, and some cherish the times when the annual flood replenishes wildlife.

The sort of experiential knowledge and practice that influences people's history, beliefs, activities and passions about water-places that contain, channel and sometimes transform fresh and saltwater sources, provides a unique trajectory to understand what socially, politically and emotionally connects people to place. This situation emerges when water sources are potentially or actually affected by unwanted change or developments that threaten, and generate friction, about its place-based cultural value. That sources of friction, and forms of attachment, can be understood as complementary presents a framework that might help to extend knowledge not only about the water/human relationship, but also about how place-based cultures are given meaning and by whom. This scenario cannot be solely confined to Kimberley indigenous groups. For some of the reasons I have described, non-indigenous peoples have also developed complex attachments to water-places overtime.

One of the implications from this discussion is to inquire whether it mattered that the sites of focus were specifically marked as water-places. I have suggested here that while the waters and rivers mattered, engagements about their protection mattered more. It was through multi-layered social and political engagements over environmental and ecological protection that place-based cultural attachments were formed.

Postscript

Contestations over Kimberley water-places and environments have been increasingly amplified in the electronic, print and cyber media. With saltwater places and nearby islands under imminent threat of gas project development, local concerns have intensified. While indigenous groups have finally secured 'grip' via native title success, many are now facing the 'irritation' that such success has engendered. Notwithstanding ecological concerns, native title leverage ensures economic support (such as work and housing) if the project is agreed to, but not if it is declined. With or without native title success, indigenous groups cannot veto such a project. Whilst in no way categorically oppositional, alongside indigenous concerns are those of a non-indigenous kind, especially for people aiming to privilege the rich biodiversity of local waters, wildlife, lands and coasts.

Lines are being increasingly drawn and intersected in human/ environment articulations that cut across social groups (indigenous, settler, migrant), disciplines (conservationists, ecologists, naturalists), industry (pastoralists, developers, tourist operators, agriculturalists), and government and non-government agencies. Not too far from each interest and claim, are social and political attachments that reveal culturally complex anchors to place.

6

Managing Water Diversion

Water is increasingly a major political issue as scarcity of the resource grips several societies, especially in developing countries where agriculture is responsible for about 80 percent of water consumptive use. However, western nations are not immune to water tensions, for instance in Greece, Spain, Italy, or the western United States, where the available water is being exploited to the limit: the Colorado no longer reaches the sea, and growing debates are emerging as to whether water should be allocated to thirsty cities or to agriculture; whether public funds should be invested again to increase the resource or demand management implemented; and whether water could be imported from far away.

These questions are increasingly relevant: consumption patterns of water in the western US are clearly not sustainable. Given the technology available today, massive water transfers could only come from Canada. There have been several mooted projects in this vein, mostly in the 1960s and 1970s. Should Canadians worry about water exports to the United Sates, especially in the frame of NAFTA? Or is climate change going to be the decisive factor in the debate?

The Advent of Water Addiction

Water is a key ingredient in the fabric of the western American society, as has been well studied by Donald Worster. The west is not completely waterscarce, for several mighty rivers flow in the region, mainly fed by snow and glaciers from the Rockies; but it definitely is a semi-arid region, compelling all societies living there before the industrial revolution to adapt to water scarcity. Early 20th century American society, empowered by the industrial age, decided to harness rivers and aquifers. Technology enabled American society to eliminate the water scarcity burden and developed the illusion that technology would always bring about a solution to growing water needs. "What nature does not yield freely, humanity should refashion to better suit human needs.... Nature has no greater purpose than to serve Man, and Man has no greater purpose than to work the land and take his place in the productive cycle."

Wendy Nelson Espeland also clearly depicted the representation that developed at the time that all water flowing unused to the ocean was a wasted resource, and that rivers needed to be "tamed" and "harnessed" to be put to use. President Franklin Roosevelt declared in 1935, when inaugurating Boulder Dam, that "the mighty waters of the Colorado were running unused to the sea. Today we translate them into a great national possession." This idea is still very much alive today. "Much of the contemporary culture of the American Southwest is based on denying its desertness.... In a subdivision being built to the south [of Las Vegas], Paseo Verde Parkway and Val Verde Road intersect in Green Valley Ranch. The concept of green, like sod lawns, was an imported fantasy."

Technology can make up for water scarcity and water must be used so as to develop western resources: "The rural Western ethic is that all wealth comes out of the ground, either as grass growing or as minerals being mined.... So the fact that today's reclamation projects-such as Garrison, CUP, Animas-La Plata-cost a few million dollars for each farmer they put on the land, doesn't cause their proponents to blink. That, they say, is the price society pays for creating the stuff of wealth. Without it and the other industries based on earth, there is nothing."

Early Massive Water Transfer Projects

Massive water transfers were first built in the eastern United States in 1847 with the Croton aqueduct for New York City and the 1900 Chicago diversion. As early as 1906, water transfers began being built in the west, with a 46 km-long canal from the St. Mary River in Montana to the Milk River, triggering a severe dispute with Canada that forced the United States to negotiate the 1909 boundary waters treaty. Other projects soon followed: the Los Angeles aqueduct in 1913, the Hetch Hetchy water supply (1934), the Colorado River aqueduct (1941), the Ail-American Canal (1942), and the California central valley project (1951). Diversions were initially designed for municipal water supply, but irrigation soon became their main rationale, as the sector benefited from generous subsidies from the federal government through the Bureau of Reclamation (created in 1902), its agency dedicated to the development of agriculture in the west.

When exploited resources began to show signs of exhaustion, engineering firms started to design huge transfer projects from Canada. But the main trigger of the idea that water resources were exhaustible and not adequate to sustain limitless development in the western United States struck in 1963 when the supreme court compelled California to stop excessive Colorado water pumping. Colorado River water had been apportioned between upper-basin and lower-basin states with the Colorado compact of 1922, but Arizona felt California tried to prevent its own use of the water. With the 1963 ruling, California's share was established at 4,4 million acrefeet (af), by far the largest share in the 15 million af available, but a figure that set a limit to what California could bank on. The very idea that the water resource was all of a sudden set at a strict limit sent shockwaves throughout California that triggered a frenzy of plans designed to increase supply through continental water diversion projects from Canada.

These projects, like the earlier water diversion schemes, involved diverting large volumes of water over long distances, but their sheer size placed them on a very different scale, both regarding their length (often greater than several thousand km) and their cost (several dozens of billions of dollars). They were designed, from 1959 to 1985, in the United States and sometimes in Canada, on a magnitude no project had ever reached before. These projects involved the transfers of the Mackenzie River, for instance, all the way through the flooding of the Rockies Valley to the Mississippi and Colorado Rivers, or the diversion of the Great Lakes, or even the damming of James Bay so as to convert it into a freshwater lake and its subsequent transfer to the American west. Costs involved reached hundreds of billion dollars and environmental impacts were not studied at the time.

Massive Continental Water Diversion Projects Never Implemented

Although these continental water diversion projects received considerable media and state government attention in the west in the 19605, none went beyond the feasibility study level and none managed to get an official approval.

Cost

Water is heavy. When transported by aqueducts over long distances, it requires a lot of energy merely to be moved and pumped over obstacles. The Grand Canal, designed in 1959 by Canadian engineer Tom Kieras would have required at least 6 nuclear plants merely to pump the water from James Bay to Georgian Bay in Lake Huron. It is therefore expensive to operate these infrastructures. Moreover, the capital requirement is huge, as attested by their estimated construction costs at the time of the project design. Governments were more than reluctant to invest such large amounts of money for water exports, especially when the profitability of such projects was far from certain.

Estimations of the cost of water transported or produced by different means show that desalination is becoming much cheaper than it used to be: about $0.6 to $1.2 per cubic meter, as compared to $0.55 to $1.35 for plastic container transportation; $1.25 to $1.5 with water-carrying ships; $0.6 to $0.85 with iceberg transportation; $0.07 to $1.8 for water recycling; and $0.8 to $3 for transfer canals of about 500 km. It is true these means do not all serve the same purposes, as desalination plants are usually located by the sea and serve urban demand. Sea water desalting technology, in particular, improved so fast between 1985 and the present that operation costs have been reduced dramatically and divided by five, from $2.5 to $0.55 per cubic meter.

Desalting is now a very affordable water-producing technology for urban and industrial consumers, enabling them to tap into an inexhaustible source. However, the water it produces cannot be considered for irrigation purposes, given its cost, its distance from the interior, and the large volumes irrigation demands. But these figures are useful inasmuch as they show that water transfer by canals is not as inexpensive as many advocates insist it is: it is not certain that farmers could afford their expensive water volumes without massive subsidies.

Stabilizing Demand in the United States

The main reason that the giant diversion schemes never came to fruition is that demand was not really present in the United States for Canadian water. Several factors explain why these projects lost their allure for both governments and society. The budget crisis that began in the early 19705 in the United States precluded public money for such large endeavours, at a time when public opinion gradually turned against large scale engineering projects that tampered with the environment: 1977 was the "hit list" year when President Jimmy Carter discarded several water development projects in the west because of their extremely low benefit/cost ratio. The economics of continental diversion projects have so far worked against them and will do so for several more years.

It was all the more difficult to argue for these continental diversion schemes as water use, contrary to what had been expected in the 1960s, showed a definite trend toward stabilization, especially in agriculture. More cost-recovery-oriented water pricing; competition from other regions, mainly Asia; and cost incentives that lure American producers to Mexico are among the factors that explain why water use in agriculture remained roughly stable between 1975 and 2000 throughout the country. If the federal government does agree, during the Doha round of trade negotiations, to reduce agricultural subsidies, water prices for farmers could increase markedly, resulting in financial incentives to consume less. Alternatively, water demand for irrigation could decrease as the result of farmers going out of business. As a whole, water withdrawals in the US have increased slowly, much more slowly now than its population, and could even begin a downward trend should competition from foreign fruit and vegetable producers work against local farmers.

In addition, although there still is room for improvement, water use per person is showing signs of stabilization, probably thanks to both tariff and education policies. The evolution of water use in the United States, from 1970 to 2000, shows that although the population grew by 38.6 percent, water use increased by only 10.7 percent. Urban consumers began to value water conservation with the development of urban water use reduction programs in most large cities. Industrial use actually decreased by 55 percent, whereas irrigation grew by only 6.1 percent.

Therefore, although water is still used at an unsustainable rate in the western United States, importing water from Canada is not as urgent as appeared to be to some politicians a few years ago, such as former senator Paul Simon. A stabilizing trend in water withdrawals and the availability of cheaper sources with desalting plants led public planners to forget about massive water transfers from Canada. When asked, even the Western States Governors Association estimated continental diversion projects were unlikely to ever be built."

Opposition from Within the United States

Moreover, there also was opposition from within. Governmental archives from the early 1980s attest to the western United States lobbying for the diversion of Great Lakes water to quench its thirst. The International Joint Commission, created to prevent and resolve disputes between the United States and Canada under the 1909 boundary waters treaty, explicitly warned against water diversions from the Great Lakes basin in its 2000 "Final report on protection of the waters of the Great Lakes."

Great Lakes states wanted to resist these projects, both for environmental and political reasons: why would the Great Lakes states give California added value at a time when so many firms were leaving the area and moving to the West Coast? The Council of Great Lakes Governors (CGLG), created in 1983, is a partnership of the governors of the eight Great Lakes states and the two Canadian provinces of Ontario and Québec. The Great Lakes charter stemmed from their growing concern that Great Lakes water could be diverted to water-scarce regions of the United States.

The charter, signed in 1985 by the CGLG members, created a notice and consultation process for Great Lakes diversions. The signatories agreed that no Great Lakes state or province would proceed with any new or increased diversion or consumptive use of Great Lakes water over five million gallons per day without notifying, consulting, and seeking the consent of all affected Great Lakes states and provinces.

After an Ontario firm, Nova Group, tried to tap water in 1998 from the Great Lakes for export purposes, triggering an uproar of protests from both the Canadian and the American sides of the lakes, the CGLG decided to further strengthen the water export ban. This led to the Great Lakes charter annex, signed in June 2001.

The annex outlines a series of principles for reviewing water withdrawals from the Great Lakes basin that is grounded in protecting, conserving, restoring, and improving the Great Lakes ecosystem. In December 2005, the annex 2001 implementing agreement was signed, with specific guidelines for the enforcement of the principles. It will also form a legal basis for the ban when the Great Lakes-St. Lawrence River basin water resources compact is ratified by the US congress, making it legally binding for American Great Lakes states. With this agreement, water diversions are only permitted over short distances from the Great Lakes, must ensure that water is returned to the Great

Lakes basin, and are only permitted if there is no other solution to satisfy the water demand.

Massive Transfers

Canadians became very fearful in the late 19905 about potential American water transfer schemes, leading to calls for the federal government to enact a ban on water exports. Wisely, Ottawa refused to enact any law that would implicitly recognize water as a good, but rather put forward bill C-6, voted in December 2002, which forbids interbasin transfers. Under NAFTA, this was a clever move inasmuch as it leaves water outside the realm of commerce and applies to both Canada and the United States. But Canadians were ignoring the fact that large-scale water transfers already exist throughout the world, and especially in Canada. Several massive transfer schemes have already taken place: in British Columbia (the Kemano transfer, 1954); in Manitoba (the Churchill derivation, 1976); in Ontario (the Long Lake and Okogi River transfers, 1939 and 1943); in Labrador (the Churchill Falls transfers, 1971); and in Québec (the Caniapiscau and Eastmain transfers, 1985).

There are major differences from American water diversion projects: Canadian transfer schemes were for hydroelectric development purposes, and not for urban consumption or irrigation. They are usually massive and over short distances, less than 200 km, whereas American water diversions often run for more than 500 km. There are currently plans to divert the Peace River in northern Alberta to meet growing agricultural and urban water needs, but that project is far from accepted. In addition, HydroQuebec is currently developing other water diversion projects for the Betsiamites, Eastmain, and Manouane rivers development.

There is controversy over the extent of the negative impact of these diversions on the environment, since the concept of minimum ecological flow remains controversial among biologists. Water does remain in Canada for each of these diversions, but this is not the point: Canadians collectively ignore the fact that their daily comfort and economic activity rest, to a certain extent, on major river diversions. It is therefore difficult to defend the argument that water transfers to the United States would weaken the environment, if one refuses to consider the phasing out of transfers within Canada.

At the end of 2005, with the signing of the annex 2001 implementing agreement, it was safe to say continental water diversions were definitely outdated as a water management concept. Demand was stabilizing in the western United States; public opinion had developed a strong distaste for the idea; governments were more than reluctant to invest the billions of dollars each scheme implied; diversion were by far not the most cost-effective way to provide for additional water volumes; and conservation and recycling were promising alternatives to the long-used supply management approach.

Toward The Return of Massive Continental Water Diversion Projects

Climate change could shake this status quo. As years go by, and as scientists as well as local governments accumulate impressive data attesting to a definite trend toward climate change, the developing weather shift could prove more challenging than the long-advocated water exports to the United States of the 19605 and 19705. The possibility that climate change could affect Canada's water quality and quantity is serious enough that Environment Canada is spending several million dollars on scientific studies to try and assess the risk. Competing uses-for urban areas, agriculture, and oil production, for instance-in regions such as Alberta where water could become scarcer, could lead to conflicts between groups that would see authorities intervene with unpopular measures such as water pricing or regulation.

Present forecasts, so far proven by observed data, show increased global precipitation, but this extra amount of water would be more than offset by increased evapotranspiration (Et) because of higher temperatures. In the already dry Okanagan Valley in British Columbia, water demand for orchards could therefore be multiplied by 12 in the summertime because of increased Et. In addition, a change in the pattern of how precipitation falls throughout the year, coupled with changes in temperature, could very much alter its availability, resulting in seasonal but dramatic supply reduction, as happened in Québec in August 2002.

In the Prairies, already less snow falls in the winter, and it melts earlier in the spring, allowing for less water in rivers in late spring and summer. Along the eastern slopes of the Canadian Rocky Mountains, glacier cover has decreased rapidly in recent years, and total cover is now approaching the lowest experienced in the past 10,000 years. Rivers in the Prairies are mainly fed in the summer by melting glaciers. As the glacial cover has decreased, so have the downstream flow volumes. This finding appears to contradict projections of the intergovernmental panel on climate change (IPCC) that warmer temperatures will cause glacial contributions to downstream flow regimes to increase in the short term. Historical stream flow data indicate that this increased flow phase has already passed, and that the basins have entered a potentially long-term trend of declining flows. The continuation of this trend would exacerbate the water shortages that are already apparent across many areas of Alberta and Saskatchewan owing to drought. Climate change is of course not merely affecting Canada, but the United States as well. Similar conclusions have already been reached with regard to the potential impact of the Rockies' disappearing snow and glacial cover on the western United States. Planning for the impact of climate change remains difficult, because building accurate models for various regions-ones that give a true picture of future available runoff-remains a daunting task for programmers. What we now have are educated guesses sustained by general models and data trends. But these trends do show that several regions in the United States, especially in the dry west, could be faced with critical water shortages, given present demand.

This might prove to be a scenario in which the United States could become interested again in Canadian water: the goal would be to prevent a social crisis triggered by conflicts between agriculture and cities for dwindling water resources. The equation for the American federal government would be the perceived financial, social, and political costs of finding new water resources, as opposed to arbitrating against agricultural water needs. No one can predict what choice officials would then make if this scenario came true. The following parameters would be significant:

- Agriculture is responsible for 80 percent of water consumption in the American west.
- A 15 percent reduction in irrigation use would satisfy the needs forecast for cities and industries."
- Agriculture does not currently pay the market price for its water, thanks to strong federal subsidies; tariff increases would result in more disciplined behaviour and encourage conservation.
- Western farmers have begun selling or transferring their water rights to cities, as illustrated by the October 2003 water deal between the Imperial Irrigation District and the Metropolitan Water District in California.

Massive water diversions are not new in North America: they were extensively used in both Canada and the United States, in the latter mainly to develop the west and to bring water to cities and fields so as to sustain an imported way of life, and in the former to develop hydropower projects. Beginning in the 19605, western American states and engineering firms began developing continental water diversion projects because of fears that continued development would bring about severe water scarcity and social disruption. These doomsday scenarios did not materialize, due to reduced economic growthespecially in agriculture, as well as more efficient water use technology and the advent of water conservation, and the huge environmental and financial costs of these projects.

However, the very idea of massive water diversions is not completely dead. Tom Kierans, the designer of the Grand Canal, is still lobbying Ottawa and Washington. More challenging is the idea that ongoing climate change could dramatically affect agriculture in the American and Canadian west. Under these conditions, what answer would governments have? Whither agriculture? Would governments favour the more profitable and water-efficient farmers or bring back to the forefront the idea of transferring large amounts of water from humid parts of the continent to drier zones? The answer probably lies in the severity of the developing weather pattern change and in the adaptation capability of the agriculture sector.

Domestic Water Supply in Rural Areas: Implications for Policy

Lack of access to safe water is at the heart of the poverty trap, especially for women and children, who suffer in terms of illness, drudgery in collection of water, and lost opportunities because of the time that water collection consumes. In rural Africa, according to the World Bank, 40 million hours are spent, each year in collecting water for domestic use and half of Africa's population is without access to safe water (Black, 1998). Recent renewed focus on poverty alleviation has resulted in increased attention to the benefits of improved water accessibility. Poverty assessment research has consistently shown that improvement in water services is a critical element in designing and implementing effective strategies for poverty alleviation. In Tanzania, experience with respect to rural water improvement has, in general, been an unmitigated disaster.

Water provision facilities have invariably fallen into disuse or disrepair because the approach used to run them failed to ensure sustainability of services. As a result, a new water sector vision has emerged based on the demand responsive approach (DRA). The major principle of the vision regards water as an economic and social good to be managed at the lowest appropriate level. The DRA recognises the inherent capacity of communities in taking greater responsibility for identifying and solving their water supply problems.

Adoption of the DRA by most water and sanitation programmes has changed the approach to evaluating their success and failure. Traditionally, water and sanitation programmes have depended on detailed blueprints to provide the basis for control and predictability. Benefit-cost analysis has been a major economic tool in evaluating these projects. However, the DRA broadens the scope of evaluation and blueprints cannot be drawn up since decisions are made jointly with communities, and problem solving is based on those partnerships (Narayan, 1995). Participatory evaluation becomes an essential tool in assisting attaining stated objectives and is an integral part of overall programme monitoring and evaluation activities. Participatory evaluation requires the utilisation of different tools and approaches in evaluating the sustainability of such projects.

This chapter is part of a follow-up participatory evaluation exercise of two community water supply programmes in Central Tanzania and involves presentation of the results on communities' willingness to pay (WTP) for improving the availability and quality of water services. WaterAid, a non-governmental organisation (NGO) based in the United Kingdom and the Lutheran World Federation through the Tanganyika Christian Refugee Service (TCRS), provide substantial support and expertise for such programmes. The former operates in the Dodoma Region while the latter operates in the Singida Region. The two regions constitute the semi-arid plateau of Central Tanzania.

It is anticipated that positive WTP indicates potential community ability to recover operation and management costs (Altaf et al., 1992). Thus WTP can be used to help ascertain the potential for fulfilling sustainability, at least from a financial viewpoint. A major objective of the study was to find out if communities studied were willing to pay an increased user fee or tariff for improving and expanding existing water services. If willing, how much (i.e. what percentage) could be added to the current user fee or tariff to achieve the required improvement? Based on the random utility hypothesis, a system of Multinomial Logit functions were developed and used to estimate individuals' willingness to pay and impact on revenue from increasing the water user fees or water tariff. The WTP was calculated by estimating the payments that would cause the respondent to be indifferent after a single unit increase in an independent variable. The estimated mean WTP and proportion of individuals who are able to pay were used to calculate the potential revenue that could be generated if the policy of increasing user fee or tariff was instituted.

In Tanzania, only 40% of the population has access to clean and safe water (MoW, 1997). Between 1965 and 1985, the Government of Tanzania (GoT) spent a substantial amount of money in developing rural-based water prefects to alleviate the water supply problem, but without notable success. Most of the investments ended in failure. Government, in collaboration with NGOs, is currently trying to revive many of these projects.

The government is providing technical assistance, while the NGOs supply initial capital requirements, overhead costs and supervision in their design and implementation. When the utility is operational, the beneficiaries (i.e. the villagers) either pay for the water from the utility or contribute an annual water fee, both of which are added to the village water fund. It is anticipated that the size of the village water fund will be sufficient to operate and expand the water utilities, based on market forces. To attain such sustainability-related goals, existence of efficient water markets is essential. Sufficient funds have to be generated to meet both overhead costs and investment requirements. Information on WTP for required improvements provides an indicator of the communities' potential capability of generating funds for such purposes.

Tanzania has sufficient surface and ground water to meet its present needs. Lack of capital and uneven availability of water across regions has limited Tanzania's capacity to utilise existing water resources to provide her population with clean and safe water. As population increases, conflicts in the allocation of water between energy, households, commercial livestock, and use in the agricultural sub-sector, increases. Weak water policies, inadequate community participation, and uncoordinated donor support has resulted in wastage or inefficiencies in the form of failed water development programmes.

After independence in 1960, donors supported large-and small-scale water development schemes. However, communities participated little in their planning and management, and also contributed little to both capital and operational/management costs. The policy of the government at independence was 'free, clean and safe water for all,' the objective being to provide clean and safe water to all villages in rural Tanzania by the year 2000 (MoW, 1997). As indicated above, this policy has now been abandoned in favour of a demand-driven water development programme. The new water policy enacted in 1997 (MoW, 1997), promotes the provision of efficient, affordable and sustainable water supplies and sanitation.

The emphasis is on community planning and management, private provision of goods, works and services relating to water supply, and public sector regulation, facilitation and environmental management. Water supply or water development programmes have to be demand-driven, and the community must be willing to participate in decentralised management and cost sharing. The Ministry of Water (MoW) has been restructured and its role has evolved from that of a water service provider to that of a regulator and a facilitator.

The new water policy underscores the importance of community participation and management in ongoing and new projects. In order to improve water supply services at the village level, the roles of MoW and other stakeholders can be summarised as follows (MoW, 1997):

- Village level. Small-scale water supply projects are to he operated and managed at the village level. Operation and management costs are to be met with funds raised within communities. The emphasis is on management by participation through formulation of village water committees that oversee and manage the utilities on behalf of community members. The communities agree upon the operational modes, with specific emphasis being paid to women's participation at all stages of water project development and management.
- District level. The water department at the district level is required to facilitate training of water managers and attendants/mechanics at the village level. The department is expected to maintain a pool of experienced technicians who will collaborate with village level mechanics in servicing and repairing established utilities. Another role of the water department at the district level is to make sure that necessary spare parts are available when needed.
- Regional level. The water department at the regional level has jurisdiction in providing guidance and making sure that government water policy and rules are adhered to. Apart from managing water supplies at the regional headquarter level, the department has to provide consulting services to districts

and villages in terms of training, and facilitating availability and distribution of spare parts.

- National level. At the national level, the MoW has responsibility for financing and managing large-scale water supply programmes (especially in large cities), to train water professionals, and to finance maintenance units at the district and regional levels. Other responsibilities are to standardise capital equipment and tools used in water development, to help facilitate availability of spare parts, and to coordinate donor and NGOs activities to make sure they follow Tanzanian water policies and local government rules and regulations.

The general objective of the water policy is to propagate the new vision of community participation and management based on the DRA. Communities arc expected to appreciate that water supply is no longer a free service provided by government, but rather government or other development agencies are available to facilitate and complement community efforts at meeting their own water supply needs. Thus, this vision is currently guiding the development of community-based water supply projects at the village level in the Dodoma and Singida Regions of Central Tanzania. Detailed descriptions and the rules pertaining to the programmes in each region can be found in Kaliba (2002) or the WaterAid website.

In both regions, financing for the water projects is thus divided among the beneficiaries, the government and donors. Villagers make cash contributions towards capital costs. They contribute time and labour, local materials and contribute to hospitality costs of visiting government staff. The initial amount of money contributed to a project by a village has been standardised and agreed by the major donors working in the regions.

The cash contribution is not currently expendable but acts as a reserve fund or safety valve for the project. The community-based water fund has to be increased using profit margins generated from user fees (Dodoma Region) or tariffs (Singida Region) when the project is operational. The government provides professional staff, an annual cash contribution towards the programme, and provides some transport and most of the construction equipment. Donor (WaterAid in the Dodoma Region and TCRS in the Singida Region) funding pays for the purchase of locally procured materials, government staff fieldwork allowances, training courses and running costs. The donors also procure all imported materials, equipment and vehicles, and employ technical and management back-up staff. In the Dodoma Region the village water utilities have emphasised deep boreholes and engine-driven pumps, and charge a tariff (i.e. a specific amount per litre of water used). In contrast, in the Singida Region, much greater emphasis is placed on shallow wells (8-10 m deep) and a user fee (i.e. a specific amount per year per household), is charged. In the Dodoma Region, between 1991 and 1999, 857 projects were implemented under the WAMMA1 project funded by WaterAid. During this study period, 29% of the projects were not functioning. For those not working, 75% were waiting to be repaired, and 25% had permanent problems (e.g. missing engines, damaged pumps or dried up waterholes).

Useful indicators of progress are the growth in the populations served and village water funds voluntarily contributed by communities. The total population served increased from 0.27 million 1994 to 1.23 million in 1999. This is an increase of about 320% over a period of six years. During the same period, the contributions to the village water funds increased, in real Tsh terms, from about 0.24 million in 1994 to about 14.87 million in 1999. Since there is no evidence of a regional increase in purchasing power, and given the low income per capita in the region (BoT, 2000), this is a good indicator of the success of community-based water development projects in the region.

During that period, TCRS constructed 49 boreholes, 465 shallow wells, and three windmills. The number of villages with water committees increased from three in 1986 to 244 in 1995. In the same period, three villages did set up water committees without the help of TCRS. The number of villages with water funds increased from zero in 1986 to 229 in 1995. Although these numbers indicate the magnitude of time and money invested in water utility projects, they of course, do not provide by themselves any explicit evidence of the potential sustainability of the projects and/or progress towards achievement of sustainability goals.

To elicit WTP information from individuals, the contingentvaluation (CV) method is used. There are three major procedures used in CV surveys, namely conjoint analysis, dichotomous choice and the payment scale approaches. The conjoint analysis procedure asks respondents to rate rather than to price alternatives. It allows for measurement of consumer preference between items with multiple attributes (Baidu-Forson et al., 1997). In the dichotomous choice approach, respondents are asked whether they would vote to change the provision of some public good at a cost of $X to themselves.

The respondents answer yes or no. For the respondent who answers yes, the experiment is repeated by varying X, until the respondent says no. After that, the respondent is asked to state the value he/she is willing to pay for the good or service if provided. For the respondent, the amount mentioned is considered the maximum WTP for that good or service. The payment scale procedure allows respondents to choose a value or price (cost) on a given scale measure. A range of values are presented to the respondents and they are asked to identify one value or price (cost) they are willing to pay or incur to purchase the good or service being evaluated. The respondent's choice from the given scale indicates the WTP for the stated good or service. Positive values indicate positive demand, and zero values indicate zero demand for the good or service.

In all cases, the surplus benefits are calculated using the parameters estimated from a model based on the type of distribution attached to WTP. In most cases, the models and type of distribution to use are dictated by the type of questions imposed and responses obtained. Other good references are those of Rosenberger and Walsh (1997) for ordinary least squares models; Kenkel and Norris (1995) for models with log-normal distribution; Steven et al. (1997) for Tobit models; and a classical work by Haneman (1984) for Probit and Logit models.

CV surveys entail three operations: designing a questionnaire, conducting a survey, and analysing the results. The validity of the estimates depends on how skilfully these operations have been done (Christe and Schwab, 1995). Donaldson et al. (1997) suggest that the use of a payment scale is more valid than the open-ended or closed-ended approach when presenting hypothetical bids. The limitation of the payment scale is constructing scaling benchmarks. This limitation can be reduced through using available market and nonmarket information to construct the scales, as suggested by Donaldson et al. (1997).

From the above review, it can be seen that there are different approaches to modelling WTP. Nevertheless, all procedures involve the respondent choosing one option from a range of other alternative services or goods based on their expectations. In order to develop models that are consistent with economic theory, the McFadden (1981) random utility hypothesis has become a standard approach to modelling WTP. As far as our study was concerned, the objective was to find out if communities are willing to pay an increased fee or tariff in order to get additional money for improvement or expansion of existing water utilities. If willing, how much (i.e. what percentage) should be added to the current user fee or tariff to achieve the required improvement? Field data were collected from 30 villages that participated in community water utility projects for a period of three years in Dodoma and Singida Regions (15 villages in each region) prior to the field surveys. Dodoma Region was divided into five clusters based on tribal distribution. From each cluster, three villages were randomly selected, differentiated according to accessibility. The results of past participatory rural appraisals (PRAs) conducted by WAMMA (some as early as the 1999) were reviewed and used in making adjustments to the questionnaire and in helping to develop pre-coded responses and socioeconomic indicators. In the Singida Region, due to transportation problems (part of the El-Nino legacy), only Singida Rural District was included. Rased on information gathered at the District and Regional Water Departments and the Regional Development Office, the district was divided into three agroecological zones. From each zone, five villages were identified to form a sample cluster. The Singida Regional Development Officer also had some PRA results that were used to fine-tune the questionnaire used in the study.

Two survey instruments were developed to gather the necessary information — the village checklist and the structured questionnaire. The village checklist solicited general information at the village level. This was done through meetings and group discussions attended by community members, government leaders, project managers, pump operators, extension agents, and water committee members. The major objective of the checklist was to get a general overview of the village projects based on group discussions. The discussion also helped to pre-test the structured questionnaire and in further adjusting the precoded responses. The questionnaire was administered to 225 individual respondents in each region. In each village, 15 respondents (at least five men and five women) were interviewed. The respondents were selected randomly from the village register and grouped based on age and social status of the individuals. Sample selection was stopped when there were enough sample respondents in each group. The two survey instruments and further information about the approach used in the study can be found in Kaliba (2002).

Regarding information on participation and management, several questions were asked. The questions related to the source of the project initiative, personal involvement in decisionmaking, labour and financial contributions, and the involvement of women in project design and implementation, emphasising participation in decision-making and informed choice. Questions were also asked regarding consumer satisfaction with the services and management of the water utility. The questions focused on: getting information on personal involvement in management; rating of the water managers' performance; knowledge of how the water management structure functioned; and the possibility of influencing change. To quantify respondent participation in project activities, and satisfaction in project performance, individual's responses were aggregated and used to calculate participation and satisfaction indexes as suggested by Sara and Katz (1998) and by the World Bank/UNDP (2000). Once again, further details on the approach used as far as this chapter is concerned can be found in Kaliba (2002).

On the WTP section, a respondent was asked if there was a need to improve the current water service. The answer was either 'yes' or 'no'. If yes, what kind of improvement was needed? The respondent was encouraged to identify only one improvement. After the respondent identified the type or choice of improvement required, she/he was asked to suggest an increase in tariff (for the Dodoma Region) or user fees (for the Singida Region) in order to cover the costs of the proposed improvement.

As explained above, the analysis was directed at identifying types of water services that need improvement and ascertaining the WTP for the improvement. In the Dodoma Region, the results concerning the water utility service can be summarized as follows: about 14% of respondents indicated that they were satisfied with the status quo (no improvement); and 81% suggested increasing pump capacity or buying a more powerful engine (thereby increasing the rate of water discharge). Thirty-three percent of respondents proposed increasing the number of water distribution points (water reticulation), and 22% proposed other improvements such as fencing, construction of new water tanks, construction of security houses, and construction of livestock watering points. The suggested tariff increase ranged from 5 to 100 Tsh/20 1. The existing tariffs averaged 20 Tsh, with a standard deviation of 9 Tsh (per 20 1 bucket of water) and on average amounted to a total of 895 Tsh/household/year.

In the Singida Region, 31% of the respondents were satisfied with the status quo, 80% wanted deeper wells that can produce more water,

29% wanted to increase the number of hand pumps in their village for water distribution purposes, and 10% indicated other types of improvement enumerated above. The suggested increase in a user fee per household per year ranged from 20 to 200 Tsh. The average user fee was 503 Tsh per year with a standard deviation of 170 Tsh. During the analysis, variables for Clusters 4 and 5 in the Dodoma Region and Cluster 3 in the Singida Region were not included in the models due to correlation problems. The likelihood ratio statistics that test for no effects was statistically significant at the 5% level in both the Dodoma and Singida Regions models.

The positive and statistically significant variables in the Dodoma Region were family size and satisfaction in the performance of project activities. These results imply a respondent residing in a relatively large family or who was satisfied in the project's performance were more likely to answer 'yes' to the question regarding the need for water service improvement. A large family implies the need for more frequent water collection trips. Improvements relating to reducing congestion at watering points (increased pump capacity or number of watering points) are likely to reduce time and effort expended in water collection. Satisfaction with reference to project performance implied both an increased demand and willingness to commit resources for improvement.

Negative and statistically significant variables were age, wealth and cash contributions. Older people and richer respondents were thus more likely to choose to maintain the status quo. In the study area, older people are less likely to be directly involved in water collection activities, while rich respondents could conceivably have access to other water sources (e.g. have their own wells) or delegate others to collect water for them. The negative sign on the cash contribution variable can possibly be attributed to 'contribution fatigue.' Respondents who contributed more during project initiation or development were more likely to say 'no' to the improvement question.

Finally, respondents in Clusters 4 and 5 were more likely to be predisposed to commit resources to improvements, because the general performance of the projects in these clusters was relatively poor (Kaliba, 2002). The underlying implications of the above results for water utility projects in the Dodoma Region are that projects whose performance creates a sense of satisfaction among consumers and a strategic water fund that is rising (so as not to create contribution fatigue) are critically important ingredients in generating community-related resources for water-related improvements. In the Singida Region, females were more willing to pay for improvement than male respondents, not surprising since they are primarily responsible for water fetching activities. The fact that this variable was not statistically significant in the Dodoma Region was unexpected. However, in the Dodoma Region water utility projects produced a more reliable water supply and there were generally fewer complaints about their performance (Kaliba, 2002). Another important and positively significant variable was the level of satisfaction concerning the performance of the project.

The Multinomial Logit model was used to model factors affecting choice of services: maintain the status quo, increase rates of water discharge (pump capacity or depth of water well or more powerful engines), increase the number of water distribution points/water reticulation, and other improvements, as explained above. In constructing the Multinomial Logit models, the parameters of status quo variables were set to zero. Again, the signs on the estimated parameters show the direction of the marginal effects on choice of improvement desired. The comparison is related to the deleted choice. As an example, for Dodoma Region, respondents who perceived that water service quality was high were more likely to say 'no' for an increase in water supply or the number of water distribution points but were more likely to say 'yes' for other improvements. The interaction between quality and proposed bids were important in making the choices or indicating the type of improvement desired.

The opposite was true for the Singida Region where the interaction between quality and bids influenced the choices negatively. There, the important factor that significantly influenced the choice of improvements was individual bids. This may be related to the technology and poorer level of services in the Singida Region. Respondents may not be willing to spend more money on failing projects. Singida Region projects will have to increase performance in terms of quantity and quality of water supplied, in order to motivate community members to contribute more resources toward project development and expansion. This was particularly important in some villages where community participation in project initiation development was low (Kaliba, 2002).

Again, the reference is the status quo bid. The value of-0.145 in the quality variable in Cluster 1 of the Dodoma Region implies an increase in the quality of water by one unit decreases the percentage of respondents who want increase in water supply by 14.5%. About 64% of respondents in Dodoma and 59% of respondents in Singida voted for an increase in water supply through increased pump capacity or number of water distribution points (water reticulation). Most villages have only one watering point and it is clear that a lot of time and effort is required for water-collection activities.

Estimated Mean Willingness to Pay

Respondents who wanted to increase water supply in Dodoma Region were willing to pay 82 Tsh per 20l of water above the existing tariff. Respondents whose choice was improving water reticulation possibilities and other types of improvement were willing to pay 29 and 5 Tsh above the existing tariffs. In the Singida Region, the analogous amounts were 91, 68 and 80 Tsh per household per year, respectively. A joint t-test failed to reject a null hypothesis that there is no gender difference on willingness to pay in both regions at the 5% level of significance. In the study area, water fetching is not a man's responsibility. However, they care. Time wasted by women and children in the search for water has a negative impact on their other activities and therefore has a compounding effect on all members of the household.

Potential Revenue from Communities

Potential revenue that can be generated for project improvements was estimated using the estimated mean willingness to pay, proportion of respondents who chose the services, and reported average daily consumption of water per household. It can be seen that there is the potential for generating much more in the way of community-derived funds in the Dodoma Region compared with the Singida Region. For the Dodoma Region, the amount that could be generated is substantial by Tanzanian standards. In fact, the amount may be more than the budget allocated by the central government to regional development programmes in all sectors of the economy. However, a note of caution is in order. Care will be needed to implement policies aimed at increasing revenue generation for water improvement purposes. Higher tariff or user fees may exacerbate inequitability of access to water services because poorer households will undoubtedly be more price-sensitive than richer households. Although there are potentially mechanisms to subsidise the poor and the elderly, such arrangements tend to alienate those affected and limit their willingness to participate and be involved in decision-making processes.

Policy Implications

Results for the WTP analysis and on proposed choices for improvement, imply that respondents were eager to improve the ease and efficiency of the water collection process. For most respondents in both regions this was explicitly articulated in the form of a desire to increase water supply and the number of water distribution points. Most villages have only one or two watering points, and it was clear that a great deal of time and effort is expended in water collection activities.

There is some potential for increasing project revenues to address such needs through increases in water tariff or water user fees. However, due to the differences in the technologies to access water, it appears that substantially more funds can potentially be generated for such purposes in Dodoma than in Singida. The water technology being used in the Singida Region is apparently not solving the problem of the dry season water scarcity, let alone the issue of water quality. A new approach and vision is needed to provide a long-lasting solution to the problem in that region. Dependency on shallow wells seems to have less potential for contributing to financial sustainability of water ulility projects in the Singida Region than the deep boreholes in the Dodoma Region.

Whatever happens, in both regions, if increases in water utility levies are planned in order to improve water services, two issues will need to he carefully considered. One is to take into account the views of women especially with respect to the location of new watering points or wells. The other is to ensure that any adjustment in water levies does not create undue hardships for the community or unduly inhibit access by the more disadvantage families/individuals in the communities (i.e. the poor or elderly).

In conclusion, project sustainability, at least from a financial viewpoint, will be largely determined by the degree to which it continues to deliver its intended benefits over a long period of time. As far as Central Tanzania is concerned, there is some potential for developing sustainable water supply programmes based, to a large extent, on a community cost recovery mechanism. In villages where there has been strong satisfaction with respect to water project performance, this chapter has demonstrated that individuals are willing to contribute more resources for improvement. It is important to note that, in these villages, community members are highly motivated and participate strongly in the daily management of the projects. An increase in community participation and changing of the water provision technology for the Singida Region is important for future project expansion and new project development.

Role of Virtual Water in Solving Water Scarcity

Water, as a source of food security, plays an essential role in ensuring sustainable food resources for a growing population. However, water scarcity is a common problem in many countries and regions. Because of the uneven distribution of water resources and population densities worldwide, water demand already exceeds water supply in nearly 80 countries that support more than 40% of the world's population (Bennett 2000). Water scarcity has become an increasing constraint to their economic development, particularly food production-the biggest water user (Hong et al. 2002). Nonrenewable fossil water has been exploited to relieve the immediate pressure of water stress; this use has caused the depletion of resources and obstruction of future economic development and food security. To combat water scarcity and increasing food production, many countries have made efforts to expand irrigation.

Despite a tremendous effort, the coverage ratio of irrigation in most of these countries remains low, simply because of their scarce water resources. The growth of food production overall has lagged and has been unstable (FAO 2001). In contrast to food production, the growth of population and food demand has been rapid in China. Meeting the water-food challenge is difficult in China because of the increasing population and water demand. In order to relieve water stress, many investment projects in China are planned and have been carried out, in order to focus on changing the uneven distribution of water resources, and to fill the gap between water supply and demand.

The role of virtual water through food trade as a solution to water scarcity is not recognised in China. Indeed, most researchers do not associate the international food trade with food security. However, they now advocate regional food trading as a solution to water stress in Northern China. These opinions are consistent with Chinese government policies of food security. The considerations of the Chinese government for food trade include domestic food security and social stability, the impact on rural employment when grain production remains the major provider of employment in rural areas, and the pressure of rural-urban migration and the social economic impacts on cities (Hong 2002).

Water Deficit and Demand

Water Scarcity in Agricultural Production

The increasing demand for food needs two basic natural resources for its production, arable land and fresh water. Future food production depends on the availability of adequate and sustainable water resources. Agriculture is responsible for about 70% of all freshwater withdrawn, and it is expected that more water will be used for irrigation as world food production increases.

An FAO study of 93 developing countries indicates that some water-scarce nations are already withdrawing water supplies faster than they can be renewed (FAO 2004). Over the years, China has maintained a population increase of 1.4% and an increase in economic development of 9% per year through using a great deal of water with low efficiency. However, agricultural irrigation is constrained by water scarcity and pollution, with 70% of food output produced on irrigated land, and 73% of the total available water used for irrigation. Furthermore, agriculture is in fierce competition with industry, municipal, living and environment water uses.

Actually, in China, agriculture competition for water is below industrial and domestic demands (Hong 1999) because the direct economic output of agricultural production is rather low. In China, every 10^4 m^3 of water feeds 4.46 persons.

North China, Inner Mongolia, Hebei, Beijing, Tianjin, Shanxi, Henan and Shandong, seven provinces, use about 168.5×10^9 m^3, which only can feed 75×10^6 persons, i.e. 23% of the total population. The demand for food by the population is met by food produced in Southern China, where farmland is scarce although water resources are relatively sufficient. Present resources meet the demand for food by the large population. Indeed, the requirement for freshwater resources is four times higher than what is normally available. The required capacity for freshwater resources is the lowest of all the resources estimated (Qu 1992).

At the national level, it is estimated that there may be enough water available, but some regions will face serious water shortages, with inefficient use and serious pollution. In northern areas, water scarcity and water pollution are serious while floods and water pollution are the main problems in southern areas. 95% of available water resources in the north have been exploited, while 6% of water resources have been used in the south. Water consumption in agriculture increased in North China where land is not a limiting factor for food production. Because water use efficiency is on average 0.4-0.5, irrigation water in the north is 7.5×10^3-12×10^3 m^3/ha, or 2-5 times higher than actual crop water demand (Liu 2000). The management of water use is expected to improve for avoiding water waste and enhancing irrigation capacity.

Water scarcity in the north has resulted in higher opportunity cost than that in the south. The huge investment in the South-North Water Transfer Project also supports this viewpoint. For water-scarce China, water use efficiency is much lower than that of developed countries. With the rapid population growth, water stress is becoming a formidable challenge. Meanwhile, when water resource load capacity reaches the limit, direct cost of enlargement of food production will quickly increase and result in native food price rises. The most plausible result is that the cost of water for agriculture becomes much higher because of competition between sectors.

Future Water Demand of Agriculture

Agricultural water consumption is the principal source of total water consumption, and is increasing with growing food demand. Irrigated area has increased, but total irrigation water quantity and withdrawal do not balance. This is due to irrigation efficiency enhancement and new irrigation technology improvements. Total water consumption has increased with unchanged available agricultural water quantity. The increasing water demands of economic development and domestic needs make additional water for agriculture more difficult to find, especially in North China. The large acreage of crop production in this area indicates a large irrigation requirement.

Because different crops have different irrigation water demands and different arable land acreages in North China (Qian et al. 2001), total irrigation water demand is $48.55 \times 10^9 m^3$; 5.00×10^9 m^3 for rice, 26.53×10^9 m^3 for wheat and 17.02×10^9 m^3 for maize. Actual irrigation efficiency in agriculture is quite low (0.4-0.5), thus, actual water demand required to support production in North China is about 97.1×10^9 tonnes. The ratio for agriculture water use accounts for about 57.63% of total water resources in North China, which irrigated only 53.12% of its total ploughed acreage. The remainder of the ploughed acreage irrigation demand, as well as other sectors' demands, intensifies water stress, influencing economic development and agricultural production. A water transfer project was selected to alleviate water pressure in North China, but in the view of the economic cost, virtual water trade through food trade may be a better solution for water deficits.

Food Demand

Population Growth

The population in China is about 1.3 billion and has been increasing. Expansion of population influences every aspect of social development. Insufficient natural (fresh water, land, forest, etc.) and productive resources (e.g. food) originate from population growth. Finite resources cannot support the huge demand of the growing population. Resources allocation ratios are very small. These problems interlace and restrict economic development and social advancement. At the same time, potential risks are concealed in the problems. The population control policy, which confines the number of children in a family, cannot restrict overall growth in the short term because the growth is based on an existing huge population. It is predicted that population growth will be negative in 2050 (Kang 1998).

The most important problem is that China must have enough food for the population, whether from food production or imports. Freshwater resource scarcity and uneven distribution strain food production, resulting in an increasing portion of domestic food supply coming from imports. Such imports, in essence, have substituted for the water that would otherwise be needed for producing the food locally. Studies of virtual water and food security in Mediterranean and Africa countries have shown that water scarcity in these countries resulted in food insecurity. China has the same food issues, and uneven distribution of water and land resources has resulted in food supply scarcity. Furthermore, there is rising competition between water needed for agriculture and water for domestic use in burgeoning cities and ever-expanding industrial developments. Consequently, today the world must act wisely to conserve, preserve and improve its water supplies (FAO 2004).

Food Demand

Food demand is a function of the size of population and consumption per capita. Rapid population growth will inevitably drive up future food demand, while per capita food consumption is likely to increase as income rises, adding an extra demand for food (Hong et al. 2002). The growth in population and foodstuff demand per capita has jointly promoted the growth of total foodstuff demand in China (Kang 1998). Foodstuff demand requires that food supply increases concomitantly. However, expansion of food production is limited by farmland and water resources; thus, the pathways of food supply enlargement are expected to change. Increasing output and food import is required to solve food scarcity, together with water resources, to meet the food demand of the population. Water resources and food trade are complex issues for society.

Crop Production and Food Imports

Crop Production and Food consumption

Cereal production in China provides the basic energy requirement directly as food and indirectly as feed for animals that are used for human consumption. The increase of grain production, including rice, wheat and maize, has been rapid since 1980. The trend in grain production up to the end of the 1990s was smooth, with only small increases from 1991 to 2000 because land productivity is approaching the limit, and no new productive technology has been created to promote increasing food production. The production of cereals, cotton, oil and sugar was either unchanged or declined, while the production of fisheries and livestock products has rapidly increased.

Before 1997 most cereal production was consumed domestically, typically rice, since then, the diet has changed and more non-cereal products are being consumed. With increasing income, meat, egg and milk demand has increased while cereal consumption has decreased. In contrast to decreasing direct cereal consumption, meat consumption has resulted in indirect cereal consumption by livestock. At present, according to the current GDP per capita, China is positioned by the World Bank in the low-middle income category. However, because of rapid economic development, per capita dietary structure indicates that life quality change is one of the most important variables that influences consumption.

Food Trade in the Past Two Decades

Food imports in China have fluctuated, especially wheat imports. This import demand is decided by domestic production capacity, and is a way to ensure food supply. Production technologies, especially water saving technology and multiple crop index improvement, have promoted the productivity of wheat in China. However, the increase in food production is still limited by water and land resources. Although irrigation efficiency has generally improved, insufficient water resources are allocated to agriculture. Food products imported into China from 1980 to 2000 were mainly in the form of wheat and maize. Net import volume has decreased gradually for wheat from 11.59×10^6 tonnes in 1995 to 0.88×10^6 tonnes in 2000, and for maize from 5.18×10^6 tonnes in 1995 to zero in 2000, because domestic supply has increased. However, in contrast to declining imports, the volume of export products has increased from 1980 to 2000 due to increasing production because of improved agricultural technologies. Growing export products include rice, cotton, pork and fishery products, in which rice export rose from 0.05×10^6 tonnes in 1995 to 3.75×10^6 tonnes in 1998. Pork exports declined from 3.24×10^6 tonnes in 1982 to 1.96×10^6 tonnes in 1999 because of increased domestic demand.

With the limited potential for additional water supply, the pursuit of water sources for the improvement of water use efficiency has shifted to high-value crops. Vegetables and fruits are typically encouraged in China as economic agricultural products and as a way of raising rural incomes. The exports of vegetables and fruit increased from 1980-2000. This shift has produced high economic returns and made possible more selections of imports for the alleviation of the intensification of water stress.

Analysis of the Role of Virtual Water

Virtual Water Implications

Allan defined 'virtual water' as the water embodied in food crops that are traded internationally. Allan (1997) termed such food imports as 'virtual imports' due to the fact that they are equivalent to a transfer of water to the importing countries. The concept is used to describe a water-scarce nation's opportunities for achieving food security by purchasing a portion of its food requirements in the international market. It compares the amount of water embodied in a crop that can be purchased internationally with the amount of water which would be required to produce that crop domestically. Virtual water is a term that links water, food and trade (Allan 2003). As the metaphor of virtual water gained currency in the mid-1990s, economists at the World Bank started to use another term the 'water, food and trade nexus' (McCalla 1997).

This combines agronomic, economic and political outlooks, with emphasis on water as the key factor of production. The agronomic component involves the opportunity cost of water, which may include production of alternative crops or use in municipal, industrial or recreational activities. The opportunity cost of water use, which is a key component of the 'virtual water' perspective, must be considered when seeking an efficient allocation of scarce water resources. The same concept applies to the water embodied in other productive activities, such as manufacturing and power generation. Some manufactured goods require a greater amount of water in the production process than other goods. Nations in which water is particularly scarce may gain by importing water-intensive manufactured goods, while exporting goods that require less water in production. Similarly, nations with limited freshwater resources may gain by generating electricity in coastal areas, using ocean water for cooling, or by importing electricity from countries with abundant hydroelectric resources.

Policy Relevance of Virtual Water

Food trade plays an important role in redistributing the water resources of the world. The role of food imports in compensating water scarcity is elaborated with a quantitative accounting of the volume of virtual water embedded in their food imports. It is suggested that water be considered an economic good (Hoekstra et al. 2003a,b), and virtual water trade be seen as an instrument to achieve food security and efficient water use (Hofwegen et ai, 2003). However, when water policy issues are discussed, further implications of virtual water can be considered. Food imports are used when policy makers care about food security, while the feasibility of food imports is judged by opportunity cost. Wichelns (2003), however, thought that the optimal trading strategy could be determined by examining the comparative advantage, not the opportunity cost. This comparative advantage includes consideration of the water resource endowment and the production technologies in two or more countries that can engage in trade.

On the other side, virtual water has geopolitical implications. Trade induces dependence between countries, and it is influenced by and has implications on tariffs. Indeed, the issue of virtual water is related to globalization, which raises concerns among many politicians and the general public. Increasing global trade implies increased interdependence of nations (Hofwegen et al. 2003). Geopolitical conditions in food trade are not ignored because food security, like water security, is a highly contentious issue that needs carefully examining in terms of the degree and method of food self-sufficiency.

Domestic political power has decided traded objectives, volumes and formations (Rang 2000). Bouwer (2000) suggested that an OFEC (Organization of Food Exporting Countries) be established with international controls, and in which the importing countries are ensured that global distribution of food will not be used as a political weapon. This study emphasizes the potential role of virtual water trade on the solution to water resource scarcity and food security.

Virtual Water Equivalents in Traded food

Food production is constrained by water scarcity; thus, importing food compensates for scarce water resources. Virtual land is also evaluated using imports divided by yields per hectare in the same year from the FAO statistics, where virtual water is the amount of imports multiplied by the global average of virtual water in every crop (Hoekstra and Hung 2003). On average, 8.9×10^6 tonnes of wheat and 0.9×10^6 tonnes of maize were imported annually between 1980 and 2000. An average of 10.20×10^9 m^3 water and 27.7×10^6 ha of land would be required to produce this imported wheat, while an average of 0.32×10^9 m^3 water and 1.6×10^6 ha land would be required to produce the imported maize. The sum of 10.52×10^9 m^3 of virtual water is equivalent to 23% of the annual average transfer of water volume for the South-North Water Transfer (SNWT) project. From 1995 to 1999, net virtual water import volume of crops in China was calculated at 101.9 × 10m, and China became the fifth in the top-30 table of net virtual water countries (Hoekstra and Hung 2003). The sum of 29.3×10^6 m^3 of virtual land is equivalent to 19% of crop arable land in 2000.

Water Compensation Through Virtual Water

Water resource scarcity constrains food production, and food supply cannot completely depend on the enlargement of food production. The basic food supply comes from food production and imports. Food import is appropriate when food price is lower than domestic production cost in the international market (Allan 1999). For virtual water embedded in the cereal imports, an aggregate ratio of 1 kg grain to 1 m of water is used. As such, the figures for cereal imports can be transformed into the same numeric volumes of virtual water in m/ per capita. Research shows that virtual water volume embodied in cereal is higher than the global average (Oki et al. 2003). This means that imports of virtual water in China should be greater than estimates in this study. The factors that influence opportunity cost of production caused by the growth of irrigation water volume to enlarge the food supply have two aspects. One is the food supply capacity including food demand, food alternatives, food productive capacity and native import capacity. The native import capacity includes international food prices, native foreign trade conditions and international aid. The other aspect is concerned with water resources, embodied in water use efficiency in agriculture, agriculture management, water resource volume, water pollution, wastewater treatment and reuse. Those limiting factors are complex and, when combined, the opportunity cost in different nations and regions are obviously different. The amount of agricultural produce derived from a certain amount of water is subject to a large number of variables. These include the geographic location of the region where crops are grown, the irrigation system used, the management of the irrigation system, soil types and climatic conditions (Turton et al. 2002).

Moving virtual water will be much cheaper than moving water itself (Bouwer 2003). Global movement of food from water-rich to water-poor countries should be just as feasible as moving petroleum products from oil-rich to oil-poor countries. Virtual water opportunities for saving local water resources include imports of electric power from areas with more abundant water for cooling of thermal power plants, or with dams for hydroelectric power production (Bouwer 2000). Currently, most researchers in China think food demand and supply in China are on firm ground. Some agreed that the food import dependence ratio might be augmented to 20% (Wang et al. 2002), while others argued that food supply should be self-dependent to ensure food security and food trade between regions within China (Cheng 2003).

Virtual water is useful to alleviate water scarcity pressure, but many governments are worried that over-dependence on other nations will result in a food crisis and social instability. Importing virtual water embedded in traded food in China can be seen to alleviate water stress, and even achieve food security. Virtual water trade may compensate for water demands of not only the past but also the future. Meanwhile, the trade can store water in its virtual water form, especially by food storage, to play a potential role in solving food problems. However, policy considerations of domestic water and food security are needed to promote better understanding of the measures that need to be taken to ensure their future supply.

7

Harvesting Water from Streams in the Semi-arid Environment

The Himalayan kingdom of Nepal is sandwiched between two giant neighbours, India and China. It is noted for its unique landscapes, all found within an area of 147 181 km. Nepal is rich in traditional technologies (TK) in all sectors of rural and environmental livelihood management. Farmers are ranked high in the rural social system (Mishra 1998) because the majority of the rural population depends on agriculture, which revolves around their indigenous practices. Nepalese river systems are composed of an enormous number of small streams that originate in the mountains. These stream systems are the backbone of rural mountain society, underlying socio-economic development of the region. Streams are the major source of water in semi-arid zones of many mountain regions, on which rural agriculture and local livelihoods depend; hence, Nepalese farmers pay high regard to these water-courses.

Streamwater harvesting, particularly for drinking water supply, irrigation and watermill operation, has been practiced in the region for 3000 years (Verma 1998) and is still widely exploited as an integral part of mountain life. Direct use of groundwater in mountain regions, through pumping or other manual methods, is relatively inconvenient, therefore sources of irrigation are generally based on rainwater and Streamwater collection. Since rain is not uniform in time and space, Streamwater is the only reliable and viable means to support local agriculture.

This chapter elucidates indigenous activities undertaken by the local farmers from three different watershed zones in Nepal, in the course of harvesting water from streams to employ in agriculture and energy generation. Study areas were selected from the mid-western region of Nepal, from three neighbouring districts, Gorkha, Tanahun and Baglung.

Local Stream as a Traditional Asset

Streams are locally called Khola and Kholsa and are formed by the accumulation of water from large numbers of natural springs scattered throughout the watershed zone. Kholas arc usually the perennial tributaries of rivers, whereas Kholsas are small perennial or non-perennial natural drainage systems that are mostly the tributaries of Kholas. Both Kholas and Kholsas have an important role in sustaining rural livelihoods. Kholsas are basically steep hillside drainage channels (average > 20°) that can provide drinking water as well as upland irrigation facilities, whereas the Kholas have a moderate slope (50-30°) and are found in valleys, wherein most irrigable agricultural lands are situated. Because of their perennial discharge into streams and their proximity to grain-producing fields, Kholas are also used for the operation of flourmills, a typical traditional waterpower unit.

Rivulets are periodically fed by monsoon rains and persistently by springs that are recharged, along with the recharge of aquifers, by seepage in monsoon seasons. The unit of measurement of water discharge is based on the arm length of an elder (1 arm = 1-21 per sec at 30°) and on the amount of work done by a pair of bullocks in one day (1 Haal = 5-10 1 per sec). The minimum flow of a Kliolsa is 0-50 1 per sec, and that of a Khola is 10-1000 1 per sec. Mountain streams get plenty of water throughout the monsoon season and scant water in the dry season. Farmers have been using traditional systems of streamwater delivery, locally called KuIo, with simple headwork and accessories.

Kulos are built and managed by farmers, with no government assistance, and based on oral codes of practice among different users. Local people maintain a large kinship network that occasionally involves meeting farm labour needs (Online Encyclopedia 2003) through sharing labour or occasionally using paid labour and volunteers. Local farmers in all three study areas have developed their own mechanism for the allocation, distribution and application of available water in an equitable manner so that every field is irrigated when required and watermills run without hampering peak irrigation demands. Since most Kulos are temporary in nature, farmers refurbish them prior to each cultivation period. Mostly, Kulos are privately owned but some belong to micro-cooperatives involving 2-5 households and a few are larger and involve 5-10 households.

In Dhundhure watershed, Gorkha, some farmers have private and cooperative Kulos but some have neither private nor cooperative Kulos, and irrigate their land using residual water from uphill fields, often without contributing to the KuIo system. Depending upon the discharge into streams and the size and shape of land, some farmers have 2-3 private Kulos. Multi-owner (cooperative) Kulos are preferable if there is sufficient discharge from a given section of a stream. Although there should be sufficient water for everyone, sometimes farmers want to have separate Kulos because of the time of their peak demands, crop patterns or conflicts with that of their neighbours.

Streamwater Harvesting Technology

Mode of use

The indigenous water technology systems are extensively used by farmers from Gorkha, Tanahun and Baglung districts, and still work without requiring modern modifications. A common way of trapping water from streams that has been adopted by local farmers consists of simple dam and canal systems. Since the level of water in streams is lower than the ground level, and in order to maintain gravity flow, Kulos are aligned across a natural hill slope, following the consequent contours of the hillside. The fundamental reason for trapping such water in all study areas is for irrigation and waterpower while a little is used as drinking water. This traditional water harvesting technology is appropriate for hilly terrain.

Technical Description

Flow-diverting structures, basically dams, are small and very temporary units, constructed by successive farmers at frequent intervals along the watercourse. Some dams, such as those in a valley, can feed water directly to the fields without requiring a canal (KuIo) but, in most cases, canals from a few to hundreds of metres long are constructed to deliver water to agricultural fields or to watermills. The centre-to-centre distance between two consecutive dams is usually fixed by rule of thumb in such a way that each dam can trap the desired quantity of water, even during the dry season.

In Gorkha district within the watershed area of Dhundhure Kholn, farmers have different types of dams and numerous canal systems eventually supply enough water for two seasonal crops: Kharif (mainly paddy) and Rabi (mainly wheat). Dhundhure Khola is a small perennial stream that borders the Khoplang and Anpipal village development committee areas. The main sources of this stream are springs scattered over a 10-km^2 catchment area. Some intermittent parts of this stream are completely dry in the dry season because of exploitation upstream, but the remaining areas maintain some stream flow from recharge by groundwater or percolation from upstream. The average dry weather flow at Suntale, a small settlement near the mouth of the stream, is about 15 1 per sec, whereas the flood season flow is about 2000 times higher. It takes about 3 h to walk to the site from the nearest road head at Terakilo on the Muglingn-Gorkha highway.

Types of Dam Structures

Local farmers from Dhundhure catchment use three forms of dam. In the upper catchment, the streambed slope is very steep and often restricts construction. The only option is to find stable rocky streambeds that will not deform and can cope with high water velocities. A Rana family from Kholli Gaun village is lucky in having a permanent natural dam to irrigate their small farm near Thulo-ATio/a (local name for the upper reach of the stream) and they do not need to spend time in dam rehabilitation. Likewise Mr. Malla from Lokendi watershed in Gumlekh village, Virkot village development committee, in Tanahun District has been able to adapt the intake in a huge rocky zone in Lokendi stream without the need for artificial diversion measures.

A second type of dam is semi-permanent in nature, consisting of big surface boulders laid in the streambed to check streamflow. Since the boulders are large in size and the stream width at the site is very small, little effort is needed to check the How and direct it into a canal. Farmers normally use clay, straw and/or brushwood or foliage to fill small gaps between the boulders to maintain flow in the canal while not depriving downstream farmers of water. This structure requires frequent maintenance. Farmers with rice fields in the middle part of the Dhundhure catchment, where the slope of the streambed is moderated, use this kind of dam.

The third type of dam is usually temporary. Farmers put small stones across a stream with or without supporting wooden pillars and cover the structure with soil sods cut from fields, which make the structures impermeable and prevent erosion. Some farmers put layers of wooden logs across a stream to facilitate diversion. These structures are suitable for use in the lower valley, where the gradient and flow velocity is less destructive. This structure can also be utilized to raise the water level in a stream and to maintain gravity flow, whereas the previous two dam types are used only for water diversion. In the lower part of the catchment, farmers get plenty of water throughout the year via extensive recharge from groundwater. Fields are in propinquity to watercourses and may thus be endangered by heavy bank cutting and inundation due to floodwater.

Intake Structure

Indigenous knowledge that has been used in the abovementioned watershed areas has largely ignored aspects of water regulation through provision of intake structures. In the dry season, flow in streams is scant and every farmer attempts to take as much water as possible. Problems also arise in the flood season when occasional flash floods flush out the dams and anterior canal structures and sometimes destroy them, with little hope of rehabilitation. Farmers are considered lucky when they have land in the big boulder zone or have rocky outcrops at the water intake site. In such cases the impacts of floods and the extent of damage can be reduced.

Likindi is a small perennial stream with a catchment area of about 20 km^2 and is a tributary of the Seti River. It borders the villages of Kahunshivapur and Baidi in the Tanahum District. The main tributaries of the Likindi are the Lokendi and Bordi Kholas. It takes about 4 h to reach the site from the nearest roadhead at Damauli. In the southern part of Tanahun District, particularly within the watershed zone of Likincli Kliola, farmers have all made short handmade tunnels at the intake points where there are rocky projections. These tunnels vary in size and shape but average of about 1-4 m long and are mostly circular in shape. The size and shape of openings are made in such a way that:

- There will not be problems in directing the desired amount of flow
- They can be simply inspected and deposited debris can easily be removed
- Excess flow cannot enter the canal
- Clearance above the opening acts as a deflection wall.

Constructing such tunnel in hard rock lies within the capacity of the local community. Locally trained and customarily experienced people, generally a group of blacksmiths, can be hired to carry out this job. The relatively large Daram Kholn watershed lies in the northern part of Baglung District. It takes about 9 h to reach the nearest town, Baglung Bazaar. The region is characterized by rolling terrain and rapidly flowing streams. Vegetation and land cover in this watershed consist of agricultural land, forest, residential and commercial areas. Water from the Daram Khola is extensively used for irrigation and hydropower generation. Studies focused on the Galkot region, near the settlements of Harichaur and Hatiya. Farmers in Baglung district also have similar underpasses into their irrigation and watermill systems.

Mr. Basnet from Kauleni Harichaur village, has adopted this system for his private watermill canal. It includes facilities to check floodwater and surface runoff triggered by excess rainfall and oversupply in the stream due to frequent releases of water by upstream farmers. Excess water enters the canal and overflows the system. In order to maintain appropriate discharge, Mr. Basnet has adopted some check tunnels in canal sections before the watermill and in an extended irrigation-canal. In some places, checks are artificially made with a spillway. Although, in the flood period, such passageways act as orifices and supply little extra discharge, this has not yet been a problem for the farmers.

Canal construction

Unlined earthen canal is best suited in traditional Nepalese practice, but some other methods are appropriate for local geomorphology, such as canal lining, rock cutting and over-crossings. In steep terrains, since the water level in streams is much lower than in the fields, canals are essential to deliver water to the fields. In such cases, very short canals are laid out in relatively flat areas, crosswise to the stream profile. Fundamentally, there are three forms for constructing canal systems:

1. Soil excavation and trench formation
2. Canal shaping in cliffs
3. Clearances of boulders
4. Raising the canal-bed by placing layers of stones in the bottom.

Forming canals by excavating soil from the ground and levelling it, by allowing water to flow over it, is a common way of constructing a canal. Work is always started from the water intake point. Farmers tend to work from intake sites with flatter gradients so that any possible upcoming obstacles blocking the flow can be bypassed through the addition of canal drops without losing the gradient required to maintain flow to the field. Selection of a dam site and the alignment of canals are usually made with reference to the other existing canals. This is why almost all canals in the Dhundhure watershed are found parallel to each other. Some experienced farmers can independently judge the alignment by walking the ground and inserting pegs at appropriate distances.

The main difficulty while constructing canals is the work involved in cutting out hard rocks or breaking up huge boulders. Local farmers have adopted several methods to tackle such problems:

1. Heating the rocks by burning wood on them for a few hours, then pouring water onto the hot boulder surface. This process is repeated until there are sufficient fissures in the rock, which make it easy to break. This method was extensively used by Mr. Basnet from Harichaur Baglung.
2. Mr. Malla from Gumlekh Tanahun has a slightly different method. During canal construction, Mr. Malla encountered boulder obstacles, particularly on steep hill slopes. He scoured the soil away from around the boulders, making a hole to maintain desired water discharge. In another site, near to the water intake where it was difficult to dig out soil from a deeply rooted boulder, he simply changed the intake site.
3. Farmers from Dhundhure watershed prefer to break boulders into pieces or form canals in cliffs using a hammer and chisel, however this takes a long time. Part of a canal (about 40 m long) through a cliff near Chanpen Kholsi in Suntale village was carved by a local man, Mr. Vhuktaman, using a locally made chisel (Chhinu) and hammer (Hathauda). Similarly, in a microhydropower project, located in Likindi watershed in Tanahun, local blacksmiths were hired to cut the canal through a 50 m-wide cliff. The canal was 0.45 m wide and 0.65 m deep and ran for 15m above the stream bank.
4. If the canal alignment is shallow or close to a stream bank, farmers usually prefer to build a platform to raise the canal bed above stream level.

Silt Traps and Canal Linings

In traditional Nepalese practice, preventing silt from entering the fields through a canal is not considered to be important. Farmers from Dhundhure watershed have never experienced such siltation problems on their farmland, saying: 'Little silt enters, but it would be useful to level the uneven surfaces of the canal bed and block some cracks.' A watermill owner, Mr. Basnet from Baglung, has a serious problem with his water turbine and silt particles. He has excavated a numbers of pockets on the relatively flat bed of the canal, about 0.1 m^3 at frequent intervals to trap the silt which is then manually removed.

Since earthen canals are porous, losses due to seepage and percolation are experienced. Minor percolation losses can be prevented by mixing clay particles in the water; these then settle and fill the cracks, thereby checking water loss. Where a canal passes through boulders or sandy zones, the problem can be solved by paving the inside of the canal with slates. Evergreen Cynodondactylon (dubo) grass is also used for inner and outer lining of canals. This grass not only reduces percolation loss but also prevents soil erosion. If a cavity forms in the canal wall, it can also be filled with pebbles, foliage and clay in layers so that the section can be made stable and impermeable.

Since canals are aligned along the contour of hillsides, there is a risk of debris or/and washout from higher regions, either due to rainfall runoff or human and animal activities. Farmers from Piughar village (Bhut KIioln watershed in Deurali, Tanahun) cover some critical portions of the canal with locally available and manually shaped stones. These stones are about 0.5-1.0 m and are also used to facilitate pedestrian crossing of the canal. In lower canal reaches, logs are put over the canal to bypass debris and other material. These techniques are best for wider canals and for steep slopes.

Over-crossing Channels

Wooden channels are widely used to provide overcrossings for natural drains or in depressions. These wooden channels are constructed from notched logs obtained from special kinds of trees that are less affected by moisture. Bamboo can also be used as a channel to deliver small quantities of water. In upper Likindi, farmers from Virkot village have found another solution, since it is difficult to make a canal in the left bank where the rice fields are located, they have used wooden channels over the Likindi to bring water to the fields from the right bank.

Indigenous knowledge is conventionally used at local community level (Gorjestani 2000) in many forms, to utilize local resources (Osti 2004). Traditional knowledge is a form of social capital for the poor and their only means of support. For this reason, the potential contribution of indigenous knowledge should be positively promoted in development processes. At present, there is little or no such organized mechanism for technology transfer or to support indigenous technologies (Bischoff 2002). Technology development should incorporate water experts and engineers (Gautam and Rana 2001) so that they can precisely contribute to the technical design of water systems.

Traditional water-harvesting technology is not only concerned with cost effectiveness but also includes culturally acceptable methods that may increase the prospects of bringing sustainability into water systems. Promotion of traditional practices in local water resource management is part of the national agenda and the national policy (NPC 2001) of Nepal.

The national policy for science and technology of Nepal has also embraced ideas on water harvesting, and has focused on the ethics of indigenous technologies as an integral part of technology transfer and research and development (RONAST 1998). However, application of these policies has not yet translated into action. In the context of promoting community-based water management, traditional streamwater harvesting techniques have immense benefits and contribute towards sustainability. Good practices in traditional water harvesting need to be identified, documented and disseminated. The advantages and disadvantages of such systems include:

1. Availability of local materials: Since all components are simple and, ideally, on site with out the need to hire equipment or import materials, time and transportation costs are saved.
2. Technology: Only locally trained semi-skilled labour is needed. There is no need to organize training for farmers to motivate them to adopt the system. It is labour-intensive and pro-poor technology.
3. Environmental impact: Environmental performance of traditional streamwater harvesting technologies is comparatively good as it is practically designed and implemented to fit the natural environment. However, some additional measures may be required.
4. Efficiency and structural performance: While considering indigenous practices in modern design, efficiency of the system is critical. Traditional system efficiency is relatively low due to heavy conveyance losses, frequent need for repair and water management.

5. Management and employment: Privately managed water delivery systems perform better than those managed by groups. Large numbers of independent water users create complexities in local water management that can be simplified by integrating systems and creating users-groups to create better understanding.
6. Scale of the system: Most traditional streamwater harvesting technologies are small scale, using a very small amovmt of water for a few hectares of land or to power a small watermill. Systems are constructed for an output discharge of about 10-100 1 per sec; beyond this limit, systems become too intricate and inefficient.
7. Suitability and social acceptance: Traditional technology is culturally accepted in semi-arid mountainous regions of Nepal, where plentiful water is available in streams. Further development and modification of technology is possible but should be organized through community participation.

The era of big water scheme is ending and a focus on community-managed water systems is gaining recognition. In order to achieve the objectives of rural development, people participation is of the utmost importance. Traditional knowledge can play vital roles in bringing effective participation and, ultimately, sustainability to water systems. Traditional knowledge has a sound base as it has been tested and practiced for many years and developed for local climate and living conditions of the people. It constitutes the best foundation for holistic and sustainable development of the region.

Rivers as Resources, as Borders

Discussions concerning natural resources in Africa often question the role of the state and nation as suitable institutional and spatial scales for management. Proposed alternatives include community-based natural resource management (CBNRM) and transboundary natural resource management (TBNRM). Support for CBNRM argues that increased local-level involvement leads to more equitable and effective management of natural resources, and that communities have the most at stake in the conservation and sustainable use of locally important resources (Li 2002). By contrast, TBNRM is premised on the idea of the ecosystem being the most appropriate scale at which to manage resources. As such, management should not be restricted by national boundaries, but should cross them as necessary.

This chapter explores questions regarding suitable scales (institutional and spatial) for the management of artisanal fisheries in southern Africa. We consider the promises and challenges of fisheries management at ecosystem and local scales in a floodplain river shared by two countries; thus, in this case, ecosystem management implies transboundary management. In particular, we examine differences in settlements, users, fishing practices and institutions on either side of the political boundary formed by the river. In turn, we consider how these differences may affect the feasibility of a 'hybrid' combination of transboundary and community-based fisheries management being considered for the fishery.

Our case study focuses on the Zambezi River as it flows between Namibia and Zambia. In order to evaluate potential differences in resource use and management norms, we conducted a comprehensive survey of fishing settlements and fishers on the Namibian and Zambian sides of the river to determine: (1) the number, age, and seasonality of settlements; (2) characteristics of fishers in settlements, including their ethnic backgrounds; (3) fishing assets and activity; (4) knowledge of rules concerning what types of fishing are allowed and where, as well as the authorities responsible for setting and implementing these rules; (5) beliefs about forms and causes of fishery-related conflict; and (6) opinions regarding if the fishery should be managed, the rationale for management, and the most appropriate management authority. In doing so, we sought to determine if differences exist between Namibian and Zambian fishing settlements, fishers and fishing practices and how such differences might impact attempts to implement hybrid CBNRM-TBNRM management in the region.

We suggest that attributes of both TBNRM and CBNRM have potential application to biologically and socially dynamic environments such as floodplain fisheries (or drylands, e.g., Haro et al. 2005). However, challenges exist when attempts to fuse transboundary and community-based management carry assumptions of a smooth nesting of homogenous and congruent users, activities and institutions within a broader management area (Young 2006). These challenges, emerge from contradictions in the spatial scales of ecosystem and local-level resources, as well as the institutional scales of transboundary and community-based management.

The first challenge is the potential mismatch between the highly dynamic movement of resources and users at the scale of the ecosystem typifying TBNRM, against the local-scale resources, users, practices and institutions characterizing CBRNM. In our case study, this tension is illustrated by results showing the majority of conflicts occur when fishers from the more populated Zambian side of the river enter Namibian waters to take advantage of the more abundant habitat and fish. In the context of hybrid management, a freer movement of users in the floodplain through transboundary management focused on the ecosystem may conflict with existing locally defined rules of tenure and access to what are considered local resources (e.g., fishing grounds).

The second challenge concerns the contrasting colonial and post-colonial histories of southern African nations. These differing histories in turn affect the degree of contiguity amongst individuals, activities, and institutions brought together under supranational arrangement such as TBNRM. While fishers sharing the floodplain in our case study have similar types of fishing assets and traditional authorities, the proportion of ethnic groups, population densities, and views regarding fisheries management differ significantly. Again, if there is not sufficient common ground amongst users and institutions in Namibia and Zambia, hybrid management risks aggravating the same types of power struggles and access conflicts that have characterized other changes in resource use in Africa (e.g., Neumann 1997). This kind of failed attempt to hybridize management could also undermine any already existing and functioning community-based regimes.

Most studies of CBNRM and TBNRM in Africa focus on terrestrial resources (Schroeder 1999; Hulme and Murphree 2001; Adams and Mulligan 2003), but the rationale for applying these management approaches is relevant for many inland fisheries on the continent. The majority of African inland fisheries are artisanal, characterized by limited and/or local-level management (Jul-Larsen et al. 2003), making CBNRM an appealing option. At the same time, existing boundaries established during Africa's colonial era frequently used rivers as reference points (Sadoff et al. 2002), making many inland fisheries transboundary for at least part of their range. The importance of fish to the region's inhabitants and the potential for a drier regional climate in the future (de Wit and Stankiewicz 2006) suggests that many livelihoods will depend on perennial water sources.

The appropriate scale of management or even if management should take place at all (e.g., Jul-Larsen et al. 2003) are therefore important issues to raise in the context of freshwater environments in the region. Our study adds to a growing body of research concerning appropriate scale in floodplain fisheries management and the role of communities and traditional authorities in artisanal fisheries management. It may also serve a more practical purpose, by informing ongoing government and NGO efforts to incorporate elements of CBNRM and TBNRM in the region. Examples include the 'Four Corners' and 'Heartlands' projects of the Worldwide Fund for Nature and the African Wildlife Foundation and the recently formalized Kavango-Zambezi Transfrontier Conservation Area, described later in this chapter.

Our chapter is organized as follows: In the first section, we review the principles of CBNRM and TBNRM in general and their application to fisheries. We also examine some of the critiques of each approach and the challenges associated with hybrid CBNRM-TBNRM management, namely, the potential mismatch between ecosystem and local scales (and associated transboundary and community-based management regimes) and the effect of contrasting colonial and post-colonial influences on settlement patterns, users, activities, and institutions in adjacent countries.

In the second section, we describe our research area, highlighting the differences and similarities between Namibians and Zambians living in the area and the institutional environments in which they function. We also describe the floodplain environment and livelihoods, demonstrating the complex historical and biophysical factors existing there. These two sections provide the context for understanding our methods and results that follow. In the discussion and conclusion sections, we situate our results in the present and emerging concepts regarding fisheries management in African floodplains.

CBNRM and TBNRM in Africa: Concepts and Practice

CBNRM in Africa emerged from a perceived failure of conventional natural resource management practiced by government agencies, which relied on prohibiting local populations from using particular resources and excluding them from decision-making processes (Hulme and Murphree 2001). New community-based approaches to management by governments and NGOs (in Africa and elsewhere) seek in principle greater involvement of inhabitants, allowing both increased access to resources and the opportunity to benefit financially from protected resources and areas.

In addition, attempts have been made to make resource management itself more participatory, by including local representation and authority in management objectives and processes. In doing so,

CBNRM has two goals: (1) to enhance conservation of wildlife, biodiversity and/or the environment; and (2) to provide economic, social, cultural and political benefits to local people participating in conservation (Adams and Hulme 2001).

While the focus of natural resource management in Africa has shifted to the level of community over the past 20 years, the scope of resource management has recently expanded across political boundaries. In southern Africa, 17 areas have been identified as potential sites for TBNRM initiatives. The key principle behind TBNRM is the concept of ecosystems, or bioregionalism, where management is defined primarily by ecological scale and function (Alexander 1990). TBNRM in theory allows for a greater geographical range of protection for certain ecological features (e.g., watersheds, forests) or migratory species (e.g., wildebeest, elephants) (Magome and Murombedzi 2003). However, the rationale for TBRNM, like the rationale for CBNRM, is not just ecological. For example, the economic importance of wildlife-based tourism in southern Africa provides additional incentives for TBNRM as a means of linking tourism 'corridors' that may run through different countries (Ramutsindela 2004).

While the logic of TBNRM argues that some resources cannot be managed at the community level (Barrow and Murphree 2001), the prevalence of 'community' in contemporary thinking about resource use and management in Africa means that proponents of TBNRM cannot overlook it. As a result, TBNRM initiatives also reference local institutions and inhabitants in the region and contain important assumptions about them. Several examples exist in southern Africa where conservation goals and means are being defined both at the local and ecosystem scale, such as the Great Limpopo Transfrontier Park and the Kgalagadi Transfrontier Park (Ramutsindela 2004).

CBNRM and TBNRM in Fisheries

Community-based management (and the related concept of co-management) is a prevalent theme in artisanal fisheries. However, existing examples of fishery-related CBNRM occur largely in coastal and small lake environments (e.g., McGrath et al. 1993), as opposed to the inland floodplain environment of this study. In contrast to floodplains, marine and lake environments often have discrete and localized fisheries, making their demarcation and enforcement by artisanal fishers and communities less challenging and the potential benefits of management more apparent.

In southern Africa, attempts to devolve management and/or promote fisher participation have experienced only modest success. Challenges identified as confounding devolution include the variable activity within a fishery, conflicts with traditional authorities and government institutions (e.g., Hara 1996; Owino 2000), and ambiguity regarding management goals (such as sustainable or equitable use) and what form intervention should take.

Fisheries policy has also embraced TBNRM through the concept of ecosystem-based management (e.g., Babcock and Pikitch 2004). However, application has been limited to marine environments, where fisheries straddle national and/or international pelagic waters (e.g., Sherman and Duda 1999). These open-ocean settings usually fall outside of typical artisanal fishing activity and management issues are related to commercial fishing. In these instances, TBNRM initiatives in fisheries have not had to consider community-scale institutions or artisanal fishing patterns. In some inland water bodies, such as the African Rift Valley lakes, or the Middle Zambezi River, states have successfully coordinated ecosystem-level research (Gheb and Sarch 2002), but attempts to decentralize fisheries management and/or increase local participation have been limited to activities within states (e.g., Nunan 2006).

Potential for Combining CBNRM and TBNRM in Floodplain Fisheries

New management arrangements combining local and ecosystem scale through CBNRM and TBNRM have considerable potential for floodplain fisheries. Floodplain environments are by nature highly dynamic and diffuse. Fish abundance and catchability varies spatially and temporally (Welcomme 1985), as do the levels of fishing effort and relative tenure that can be asserted. CBNRM offers the institutional flexibility to respond to such variation, for example, to address issues of sustainability and equity during periods of fish breeding or limited food security.

Ecosystem management of floodplain fisheries is also important, given the dispersed nature of the fishery and the fact that key influences on fishery productivity, namely, the timing and duration of flooding, is an outcome of factors outside of the immediate watershed (Sneddon and Fox 2006). Moreover, our limited understanding of floodplain ecology (Arthington et al. forthcoming) would improve with data collected at the local level on catch-effort patterns in a multispecies fishery (e.g., Ticheler et al. 1998) and the integration of broader ecosystem-level trends in productivity and exploitation (e.g., Halls et al. 2006).

In spite of the potential usefulness of a hybrid approach in floodplain fisheries, we did not find any operating case studies in the literature surveyed. In cases where ecosystem-based management has been proposed for floodplain environments, policy makers typically envision a nested order of authority, from community-level management of adjacent waters, to a broader, catchment-level administration (e.g., Hoggarth et al. 1999). With this vision comes the optimistic assumption that fishers and communities have institutions compatible with this hierarchical framework and sufficient incentives to cooperate for local and ecosystem-wide benefit.

Challenges to Hybrid CBNRM-TBNRM Approaches

Questions about whether resource management at local and ecosystem levels can be combined have not been raised to date in the course of CBNRM and TBNRM co-evolution in southern Africa. This is likely because most of the areas under consideration for TBNRM already have some form of protected status. Exclusion and marginalization of local users has been a common characteristic of these areas (Schroeder 1999) and attempts to promote local involvement do not typically involve devolution of any kind of management responsibilities. Instead, focus has been directed at a broader distribution of monetary benefits from conservation or the development of alternative livelihoods not dependent on wildlife consumption (Adams and Hulme 2001). Thus, community participation in these examples is mostly at low levels (Barrow and Murphree 2001). While protected areas remain sites of contestation between inhabitants, NGOs, and the state over access, use, and benefit, TBNRM governance remains at the level of state-state negotiation.

However, the popularity of TBNRM in management discourse makes it likely that it will be considered beyond protected areas. Common property resources, such as freshwater fisheries, have less of a history of explicit conservation or marginalization, making issues of local-scale resource use and governance difficult to overlook. In the following paragraphs, we identify two main challenges in harmonizing CBNRM and TBNRM in such cases: mismatches in real (and perceived) institutional scales of management and spatial scales of resources; and the heterogeneity of users and institutions arising from colonial and post-colonial histories.

Institutional and Spatial Scales

Both CBNRM and TBNRM approaches often contain oversimplified assumptions about the structure and behaviour of inhabitants and resources of an area, and these can impact on the success of hybrid CBNRM-TBNRM arrangements. For example, CBNRM has been critiqued for its failure to adequately consider the meaning of 'community', too often considering communities as homogenous entities acting collectively to achieve common environmental goals and giving little consideration to differences within them. CBNRM often includes a process of formalizing institutions and rules, linked to identifiable communities (Allison and Ellis 2001).

This formalization overlooks the fact that most local-scale resource access and use involves ambiguous institutional borders and overlapping rationales for legitimacy (Neumann 1997; Robbins 2004). Emphasizing exclusive use and tenure to residents can conflict with seasonal resources use defining many rural livelihoods in Africa. With regards to resources, CBNRM often assumes local-scale resources and environments are relatively stable and distinct (and thus amenable to local level management), but this view has been challenged by new understandings of the spatial and temporal dynamics of ecological systems (Leach et al. 1997).

In the case of TBNRM, similarly problematic assumptions are evident. For example, while ecosystems are portrayed as 'naturally' delineated by physical and biological processes, many authors have demonstrated how TBNRM in southern Africa is influenced by regional and international political processes (e.g., Wolmer 2003; Duffy 2006). As a result, ecosystems are 'constructed' according to how certain actors perceive nature as it is or should be (Robbins 2004). Demarcation and appropriate uses of transboundary areas may be driven by narrow definitions of ecosystems, such as the behaviour of charismatic megafauna and their habitat (due to their appeal to tourism).

TBNRM proponents often also assert that as ecological networks are reopened, so too will previously fractured social networks. Wolmer (2003, 262) argues that the bioregionalism underpinning TBNRM extends to assumptions about local and regional cultures and their rootedness in the landscape. The Peace Parks Foundation (1999), the largest TBNRM initiative in the region, aspires to 'stitch' together ecological and social landscape separated by local borders (cited in Ramutsindela 2004). In practice, movement by inhabitants may become even more restricted under TBNRM. McDermott-Hughes states that establishment of the Great Limpopo TCFA implicitly carries two 'geographical notions in polar opposition: "Africa" for tourists; and "community" for peasants' (2005, 2). Small-scale resource use and benefits typical of CBNRM may continue under limited terms, but free movement encouraged for tourists is not extended to inhabitants.

Colonial and post-colonial Differences

TBNRM in Africa seeks to stitch back together ecosystems fractured by the establishment and persistence of colonial-era borders. CBNRM addresses policies towards resource use and users practiced within states. Yet in asserting either ecological or community legitimacy, the scale and influence of the state is not removed. State policies of resource use and management, as well as the identities created for (and asserted by) inhabitants and institutions persist despite changes in scale. Moreover, colonial and postcolonial policies and identities are not uniform. Even adjacent countries that were part of the same colonizing power often experienced vastly diverse trajectories. These colonial and postcolonial differences must be considered when attempting to bring together contrasting uses, management and institutions.

Adams (2003) suggests the 'colonial approach' towards resource conservation in Africa cannot be generalized. Perceptions of nature, conservation and the proper use of resources varied over time and amongst actors within colonies. These varying colonial-era ideologies and institutions persisted in newly independent African nations (Adams 2003). At the same rime, states adopted, responded to, or were coerced by the forces of socialization, economic liberalization, proxy wars and global environmentalism. These regional and global changes in political and economic structures affected each state differently over rime, adding to contemporary distinctions.

Another enduring aspect of colonial rule is how ethnic group identity and institutions were defined, created and used by administrators. Traditional authorities were co-opted or created to provide labour, collect taxes and govern certain aspects of indigenous affairs, for a cash-strapped and thinly spread colonial infrastructure (Berry 1992). Murombedzi (2003, 141) points out that 'starting with the colonial project of indirect rule, traditional authority in southern Africa has been engineered and reengineered so many times that it is hardly fair to refer to the contemporary authority figures as "traditional" at all'. Not only are communities and institutions 'imagined' (Brosius et al. 1998) but these imaginations are a product of diverse colonial and post-colonial experiences (e.g., West and Kloeck-Jenson 1999).

In the context of hybrid management, the (tenuous) assumption that TBNRM brings together harmonious communities and analogous institutions may lead to more, rather than less conflict. Scalar differences in resource use and management may aggravate conflicts arising over access to different areas by users deemed as 'outsiders' (Nielsen et al. 2004). If TBNRM fails to consider these issues, it may be destined to make the same erroneous assumptions of homogeneity and strict tenure seen in some CBNRM projects, only at larger scales and with wider impacts.

As the following description of our study area illustrates, the Upper Zambezi River floodplains have a distinct colonial and post-colonial political geography. However, there is also some degree of contiguity in cultural identity and the structure of traditional authorities. Rules about tenure and use of fisheries are also flexible and context-specific. These conditions, we suggest, present hope for a workable attempt at hybrid management. At the same time, we highlight the need to determine in greater detail where similarities and differences exist regarding the distribution of fishers, fishing activity, management and conflict.

The Upper Zambezi River

The study focused on a 120 km stretch of the Upper Zambezi River shared by Namibia and Zambia. Many inhabitants on both sides of the river are members of the Lozi ethnic group, sharing common cultural characteristics and are often related by family. However, the complex political history of the area and its high level of natural variability have acted as important factors in shaping resource access and governance.

Once part of a larger paramount Lozi Kingdom, the area was cut into two colonies during the 'Scramble for Africa' in the late 1800s (Flint 2003). From this stage onwards, the institutional landscape on either side of the river diverged. The British colonial administration north of the Zambezi River (Northern Rhodesia; present day Zambia) provided a considerable role for traditional authorities to set and enforce resource management policies (Bell-Cross 1974), although the influence of traditional authorities declined considerably after Zambian independence in 1964 (Flint 2003). In contrast, administration south of the river (South West Africa; present day Namibia) was traded between authorities no less than five times (between Germany, Britain, modern-day Botswana and South Africa) until Namibia's independence in 1990. These changes were often accompanied by tightening or loosening of access to the floodplains on the southern bank by inhabitants of the northern bank (Abbott 2005). These restrictions on border crossing arose not only due to official policy of one or both countries; during the 1970s and 1980s, the stretch of river was also a focal point of the conflict between South Africa and neighbouring states of Angola, Zambia and Zimbabwe, and movement in the area was likely curtailed for fear of personal safety.

At the national level, the two countries contrast with respect to economic and social development achievements, with Namibia ranked at 125 in the Human Development Index (HDI) and Zambia at 166. Namibia's estimated GDP/capita is US$ 6,184, compared with Zambia's GDP/capita of US$ 877 (UNDP 2005). The extent to which these national differences are relevant in our case study area is questionable, however, as the region of Namibia where our research took place has consistently had the lowest HDI ranking within the country (UNDP 2005) and Namibian residents of the study area may have income levels and abilities to acquire assets that are more similar to their Zambian neighbours than to their fellow Namibians.

Traditional authorities on both sides of the river share an almost identical hierarchical structure, with village, ward and regional representatives (indunas) and councils (khutas). All of the traditional authorities, including those in Namibia, ostensibly fall under a paramount chief (Litunga) based in Zambia. Traditional authorities in Namibia can be characterized as relevant in that many civil disputes are still arbitrated within the khuta framework (Purvis 2002). Moreover, government agencies often rely on traditional authorities as a point of contact for public awareness and aid distribution (J. Abbott personal observation). Less information is available regarding the status of traditional authorities on the Zambian side, although the limited government presence, combined with the history of semi-autonomy prior to independence, suggests that traditional authorities and structures are still relevant in daily life. Traditional authorities on both sides of the river have an ambiguous role in present and emerging fisheries management.

The Upper Zambezi River is a relatively unmodified watercourse compared to other rivers of its size, with no barriers or irrigation, a rural population and limited pollution. Annual cycles of flooding and precipitation cause a seasonal contraction and expansion of the floodplain, which in turn affects the productivity of the fishery. As areas become covered in water, fish migrate onto the floodplain to feed and reproduce. When waters begin to recede, adult and juvenile fish return to the main channel or become trapped in isolated bodies of water on the floodplain. Fishing activity is mostly artisanal, with the majority of fishers using monofilament gril nets and dugout canoes (Purvis 2002; Noesje et al. 2003). In many floodplain environments, fishing is usually not targeted towards specific species (Welcomme 1985), a pattern also seen in the Upper Zambezi River fishery; over 50 species of fish are caught by inhabitants, with the majority of individuals being either cichlids or catfish (Purvis 2002). Changes in fishing technique and effort are linked to seasonal variations in overall fish biomass and movement of fish between the main channel and floodplain (Purvis 2002).

Several sources suggest that the recent increase in the area's human population, combined with the adoption of more extractive fishing methods and an apparent erosion of traditional management, have resulted in unsustainable levels of fishing. If catches have declined, it may indeed be a result of fishing pressure, but may also be linked to reduced decadal flood volumes, changes in herbivore-mediated floodplain morphology and nutrient cycling, or a combination of all three (Abbott 2005). In contrast to this portrayal of declining fisheries, unpublished biological surveys indicate that the trophic structure and biomass of the fishery does not resemble what is expected for heavily exploited fisheries and unpublished data collected by J. Abbott during 2002 indicated a wide range of fish species and sizes available in local markets. Given these conflicting views and the limited biological and fishing effort data existing for the area, it is difficult to make concrete statements about the condition of the fish stock or the effects of present fishing activity.

At the time of the survey in 2002, the Zambian government had fisheries legislation at the national level, with minimum mesh size of 51 mm, an annual closed season between December and February and a prohibition on 'bashing' (a method of catching fish where the water is struck in order to scare fish into nets). Namibia had a very vague pre-independence legislation with traditional authorities principally responsible for fisheries management in 'Native' areas such as the floodplains (Abbott 2001). Generally speaking, the use of active gears such as mosquito nets and dragnets, as well as small mesh nets was prohibited in Namibia. However, there was and continues to be a heterogeneous interpretation of customary law both within and between Namibia's traditional authorities (Purvis et al. 2003).

A more informal, yet equally important aspect of management on both sides of the river is the silalanda, an area of land and water within a traditional authority's territory (silalo), associated with a specific kin group. Non-kin wishing to graze hvestock or fish in a silalanda are expected to ask permission (Purvis et al. 2003).

The area's physical geography has important implications for the distribution of fishing areas, settlement, and assertion of fishing tenure. The northern bank of the river is higher than the southern bank, where the majority of the floodplains are located. As a result, the majority of seasonal inundation occurs in Namibia. Floodplain settlements are therefore largely seasonal, with tenure to specific areas linked to kinship and a tradition of use over time. The gentle gradient of the floodplain leads to subtle and dramatic changes in the areas covered in water and hence the availability of potential fishing grounds, as adult fish migrate to the inundated floodplain to spawn and feed (Purvis 2002).

Recent changes in Namibian and Zambian legislation, regional policy initiatives, and NGO involvement have important implications for the current limited and ambiguous nature of fisheries management in the study area. At the regional level, the Southern African Development Community (SADC), a state-level coordinating body, passed the 2001 Protocol on Fisheries, obliging parties (including Namibia and Zambia) to adapt or change fisheries policy to incorporate traditional authorities and participatory management (SADC 2001). Also notable is the States' obligation to 'adopt equitable arrangements whereby... fishers who are traditionally part of a transboundary fishery may continue to fish and trade in goods and services' (SADC 2001). Thus, the policy recognizes the relevance of both community-based and transboundary management.

Nationally, Namibia introduced a new Inland Fisheries Act in 2003. In this Act, control of the fisheries rests with the state. However, there is some scope for community involvement (largely through traditional authorities) in the formation and enforcement of regulations, as well as coordination with neighbouring states in the management of shared waterways. Zambia is also in the process of adopting legislation with similar provisions for potential community and transboundary involvement. At the same time, the area falls within the recently formalized plans for the Kavango-Zambezi Transfrontier Conservation Area, a Peace Parks initiative involving watersheds shared by rive countries (Peace Parks 2006). The terms of reference for the feasibility study tendered by Peace Parks, outlines some of the envisioned rationales and benefits of TBNRM for the area, including 'harmonizing the policies, strategies, and practices of conserving and managing the resources that the rive countries share through natural movements' (Peace Parks 2006).

While fisheries are not explicitly mentioned in the above passage, their role in the area's livelihoods and tourism makes fisheries use and management an issue. Peace Parks does acknowledge the 'people of various nationalities, cultural backgrounds, and income levels whose livelihood depend on the shared common waters' (Peace Parks 2006). However, the goal to 'sustain the economic developments of the people' (Peace Parks 2006), while suitable in theory, risks an implicit assumption of homogeneity, underlining the need to understand where differences lie at this early stage in policy development.

Survey Approach

A 120 km stretch of the Zambezi River was surveyed, collecting multiple types of data concerning settlements, livelihoods, and management. Data collection was based on frame surveys (e.g., Moses et al. 2002), where the shoreline is divided into sections, or frames.

Typically, frame surveys are standardized so areas can be revisited after a period of time to determine if change has occurred. It is an exhaustive survey of all fishing settlements, fishers and fishing assets found within the frame. Due in part to high mobility of artisanal fishers and their assets (Allison and Ellis 2001) as well as the ephemeral nature of some floodplain settlements (Sarch and Birkett 2000), there is always the possibility our survey missed some fishers or settlements.

A previous frame survey of the Zambian side of the Zambezi River had been carried out in 1996 by the Zambian Department of Fisheries (DoF). DoF staff experienced in frame surveys cooperated in method development and survey implementation in the present study. The frame survey took place during October and November 2002, as the flood reached its lowest point.

This time was chosen on the assumption that it coincided with the peak of bank side settlement and fishing activity, since fishing yields are relatively high and the labour bottlenecks due to ploughing, which take place with the onset of the seasonal rains, are not yet apparent. Survey teams travelled the Zambezi River by boat. Given the range of flooding as well as the high banks sometimes encountered, we determined that only visiting villages in line of sight of the boat would result in inadequate coverage. As a result, all villages within 500 m of the Zambian side or 1 km of the Namibian side of the main channel or navigable waterway were visited. Different distances for each country were chosen due to the different profile of the floodplain on each side; while the floodplains on the Zambian side are less extensive due to a gradually rising landscape, the gradient on the Namibian side is almost nonexistent, allowing for greater potential inundation.

Initial contact with a settlement was made through the village induna (village-level traditional authority), to whom the goals and practical details of the frame survey were explained and from whom consent to work in the village was obtained. For larger settlements (more than five households), it was proposed that, where possible, appointments would be made the next day, allowing the induna to assemble those fishers who consented to being interviewed. Visits to settlements were made between 8 A.M. and 3 P.M., SO as not to interfere during times in the early morning and evening when fishers usually attend to their nets. Once the forms and a recording protocol had been agreed upon by the survey team, the method was tested in a village on the Namibian side and refined.

The final questionnaire had three parts: Part A contained questions about the settlement and were directed towards the induna, whereas questions in Parts B and C, regarding fishing activity and management, were directed towards individual fishers. While the final survey included 43 questions, those of interest in this chapter addressed: age, location, estimated number of households, fishing assets and status (i.e., seasonal or permanent) of settlements (Part A).

Questions directed at fishers gathered data on their age, gender and ethnic group, as well as the number and type of fishing gear (and, where relevant, mesh size) used, knowledge of current fishery-related rules and authority, opinions about sources of conflict and about what form (if any) future fisheries management should take (Parts B and C). Questions about settlements, fisheries activity, conflict and present management were open-ended, with the most frequent responses pre-coded for ease of recording. Questions about future management were in the form of discrete choices.

The eight-person survey team consisted of the lead author, representatives from the Zambian Department of Fisheries, the Namibian Ministry of Fisheries and Marine Resources and staff from a WWF-LIFE funded project titled 'Shared Resource Management on the Zambezi/Chobe Systems in Northeast Namibia: Current Practices and Future Opportunities'. A total of 72 settlements were recorded and 541 fishers were interviewed during the survey.

Selected Characteristics of Fishing Along the Upper Zambezi River

Our first set of findings concern the demographic patterns of settlements along the river, especially the distribution, size, permanence and age of settlements. Settlements were found on both sides of the river along the whole length of the survey area. We recorded 18 settlements (25 percent) on the Namibian side of the river, while 54 (75 percent) settlements were on the Zambian side. Settlement population was estimated by recording the number of households per site. A household in this context was defined as a collection of people who regularly eat together. Hence, there could be several individual structures within a single household with several related and unrelated individuals living within them.

A total of 1,709 households were enumerated, with 1,643 households on the Zambian side and 66 households on the Namibian side. Zambian settlements were on average denser (30 households per settlement) in comparison to Namibian settlements (average 3.6 households per settlement). Settlements varied according to how many months they were occupied.

According to village indunas, most Namibian settlements were considered seasonal (61 percent), while most Zambian settlements were considered permanent (61 percent). Most Namibian households were in seasonal settlements, while most Zambian households were in permanent settlements and the difference in distribution is statistically significant. However, the majority of both Namibian settlements (69 percent) and Zambian settlements (67 percent) were occupied for at least nine months of the year.

The reported age of settlements varied from one year to up to nine decades. Almost half (41 percent) of all settlements had been established within the last 20 years. The majority of Namibian (61 percent) settlements were less than 20 years old, while the majority of Zambian (61 percent) were between 30 and 90 years old. Fisher characteristics and assets: Of the 541 fishers interviewed, 98 (18 percent) were Namibian and 443 (82 percent) were Zambian. Only four of the fishers interviewed were female.

It is important to note that our results do not reflect the total number of people who may fish in the survey area. Women and children who reside in permanent settlements often use hook and line or mosquito nets to catch fish as an important supplement to their diet. While their exclusion may be potentially problematic in terms of understanding fisheries livelihoods and the impact of management interventions on local inhabitants, the research focused on fishers using dugout canoes and nets for three reasons: First, the types of artisanal fishers recorded in our study are those normally targeted by management interventions like TBNRM and CBNRM. Second, fishing with mosquito nets are most frequently practiced when water is draining out of the floodplains and illegal and controversial; hence users would be unlikely to reveal their fishing practices to our survey team. Finally, catches from mosquito nets, along with hook and line fishing, are very small compared to gill and dragnet fishing (Noesje et al. 2003).

The fishers inhabiting the settlements came from a broad range of ethnic backgrounds, with 11 different ethnic groups present in the region. The most common ethnic groups overall were Lozi and Subia, accounting for 61 percent and 20 percent of fisher affiliations overall. Namibian settlements were overall much less diverse, with only four ethnic groups recorded and over 90 percent of respondents belonging to one of the two major ethnic groups. In Zambia, 11 ethnic groups were recorded and 78 percent of fishers belonged to the Lozi or Subia ethnic groups.

Fishing assets recorded consisted of mokoros (dugout canoes) and fishing nets. While there is a statistically significant difference in the numbers of Namibians and Zambians identifying different assets, the majority of fishers in both instances own both a boat and a net. Proportionately more Namibians owned both, while a higher percentage of Zambians owned either just a boat or just a net.

The most common fishing gear used by fishers were nets. Nets can be distinguished between gill nets and dragnets. Gril nets are typically set overnight for passive capture of fish, whereas dragnets are actively pulled through the water. While there is more gear on the Zambian side (reflecting the larger population), most fishers on both sides of the river use gill nets and there is no statistically significant difference in the numbers of Namibian and Zambian fishers using gill or dragnets. Both groups used similar mesh sizes. The majority of nets reported by fishers had a mesh size between 63 and 76 mm.

Rules and authority: National law in both countries does not distinguish access and tenure between different types of fresh water bodies. However, customary law in the Upper Zambezi floodplains does provide for kin-linked tenure of permanent and ephemeral ponds and waterways (silalanda) as described above. We therefore wanted to know how tenure and access might vary throughout the survey area. Of those fishers interviewed, only 16 percent of Namibian fishers and 17 percent of Zambian fishers indicated that permission was needed to fish. Thus, most fishers on both sides of the river see the fishery as predominantly open access. Those who responded that permission was needed to fish identified specific bodies of water where permission was required, with Namibian fishers identifying mulapos (small seasonal or permanent lakes) most frequently (62 percent) and Zambian fishers identifying streams most frequently (39 percent). There was a significant difference in Namibian and Zambian fisher responses to this question.

A total of 60 percent of Namibian fishers and 32 percent of Zambian fishers said that no fishing methods were prohibited (statistically significant difference, [chi square] = 27.72, $p < 0.0001$). Among the remaining fishers, small mesh nets (note that the specific minimum mesh size was not asked) were the most frequently cited as banned by Namibians versus dragnets by Zambians. The differences in responses by Namibian and Zambian fishers are statistically significant, although the conditions for a [chi square] test are violated since no Namibians identified closed seasons as a management measure.

One method does stand out; while 20 percent of Zambian identified close seasons as a management method, no Namibians did so. Fishers who cited certain methods as being prohibited or management measures were also asked if illegal fishing took place and 59 percent of Namibians and 56 percent of Zambians agreed that it did (no significant difference, [chi square] = 0.1275, $p = 0.7210$).

Fishers who cited management restrictions were also asked who was responsible for making and enforcing restrictions. Of these, 62 percent of Zambian fishers said that the government was responsible for restricting methods and enforcing rules, while 54 percent of

Namibian fishers identified traditional authorities as responsible and the differences in responses are statistically significant.

Perceptions of conflict: When asked if fisher-related conflicts occurred in their area, less than a fifth of Namibians (18 percent) and just over a quarter (28 percent) of Zambian fishers responded that they did (no significant difference in perceptions of whether or not conflict exists, [chi square] = 3.678, p = 0.06), suggesting that conflict is not a major issue for the majority of fishers on both sides of the river. Among those who did perceive conflict, Namibians identified Zambians as the greatest source of conflict, followed by fellow Namibians in general and their immediate neighbours. Zambians identified other Zambians and their immediate neighbours as the largest sources of conflict, while few Zambians identified Namibians as sources of conflict. Namibians perceived fishing without permission as the greatest source of conflict, while Zambians saw sources of conflict spread equally between access, fishing practices and too many nets, but these differences were not statistically significant.

Opinions on management: When asked, 49 percent of Namibians and 79 percent of Zambians said that the fishery should be managed and the difference is significant. Of Namibians and Zambians who supported management, the majority of both groups identified their respective national government as being the most suitable to manage fisheries. Following this, Zambians then preferred co-management by governments and traditional authorities, while traditional authorities were identified by 25 percent of Namibians and 9 percent of Zambians as most suitable. The differences in responses by Namibian and Zambian fishers regarding management regimes are statistically significant.

The purpose of fishing regulations was regarded by most fishers of either group to be for 'conserving fish' (85 percent of Namibians and 93 percent of Zambians), with very few identifying 'keeping outsiders away' as a reason for regulation. There were no significant differences according to nationality regarding purpose of management. We begin by first summarizing and discussing our results. Then, we consider what implications our results have for hybrid CBNRM-TBNRM fisheries management in the region.

There has been a change in the numbers of settlements on the river over the course of the last twenty years, with over a third of all riverside settlements and almost two-thirds of Namibian settlements, having been established in the last twenty years. These findings support anecdotal accounts from Namibian fishers that numbers of fishing settlements, fishers and fishing activity have increased over time, trends that may reflect endogenous population growth (Mendelsohn and Roberts 1997) and perhaps an increased importance of fishing for some households (Abbott 2005). While we do not have reliable historical data on fishing exploitation to compare against present levels, we assume that as settlements along the river have increased, so has fishing activity.

Other political, historic, and geographic factors have likely contributed to settlement patterns. Politically, the end of South Africa's occupation of Namibia in 1990 led to a cessation of border hostilities as well as increased freedom of movement for Namibians. Historically, the larger overall number of fishers and settlements on the Zambian side of the river reflects the long occupation and active commerce on the northern bank of the Zambezi River in both pre-colonial and colonial eras (Gluckmann 1968). By contrast, during colonial and pre-colonial times the Namibian side of the river was used largely as a source of fish, reeds or grazing, with relatively few people actually living along the river. This settlement pattern was and is linked to geography, as the Namibian side's higher likelihood of inundation makes settlement more difficult and explains why the majority of Namibian settlements are seasonal rather than permanent.

The ethnic diversity in Zambian and Namibian settlements is an outcome of pre-colonial ethnic distributions as well as the colonial and postcolonial institutions that maintained them. The area of Zambia covered in our survey is a zone of convergence for several different ethnic affinities and this reflects the long history of commerce in the region. In contrast, the Namibian area is mainly composed of two major ethnic groups that historically dominated in the area. While most s living along the river have relatives on the Zambian side of the river, border restrictions during the South African occupation of Namibia and particularly during the bush war of the 1970s and 1980s, prevented cross-border migration. In addition, historical reports and recent events (such as the attempted secession of the Caprivi Region in 1999) suggest that inhabitants of the Namibian part of the floodplains are more resistant to migration or union with other areas (Stanley 2002).

The similarity in the types of fishing assets on both sides of the river, as well as the similarity in mesh sizes used by Namibian and Zambian fishers reflects the biological and social nature of a floodplain fishery. Given the high variability of the resource, it makes more sense for participants in a fishery, regardless of their income, to invest in very simple assets, such as monofilament nets and dugout canoes with investments going into increasing the number of units, rather than technology. In this way, participants in the fishery can more easily enter or leave as opportunity costs change, due in part to the seasonality of floodplain fisheries.

While the types of fishing assets used are similar, our survey found a significant difference in the proportion of asset ownership. Namibian fishers were more likely to own both boats and nets. At the national scale, Namibian fishers are part of a stronger economy and currency than Zambia, as well as a formalized system of social pensions (Devereux 2001). Namibian fishers may therefore have greater potential access to income that can in turn be invested in individual fishing assets.

Gill nets were the predominant fishing gear used throughout the survey area. While the difference in the use of gill versus dragnets was statistically insignificant, a higher proportion of Zambians used dragnets and it is worth considering these differences in the context of the overall size of the population of fishers in Namibia and Zambia. Gill netting is a passive fishing method that can be done by one fisher in a dugout, while dragnetting is an active method, requiring teams of two or more fishers. The higher proportion of dragnet use amongst Zambian fishers may be related to the apparent lower levels of individual net ownership, favouring group activity. Similarly, the higher ownership of both nets and boats by Namibians may be out of necessity, as there are not enough fishers to work cooperatively. However, since our survey did not include questions about collaborative fishing activity, we can not confirm this supposition. Further research on the social norms of fishing is required.

A low number of respondents claimed that permission is needed to fish and almost one third of Zambians and two thirds of Namibians reported that no fishing methods were prohibited. These findings support the claim by Scudder and Conelly (1985) as well as Bene et al. (2003) that most floodplain fisheries have low levels of management. The differences in areas identified as needing permission for access, i.e., the much higher percentage of Namibian's claiming permission is needed to access mulapos, may reflect the physical differences of each side: the more floodplain-dominated topography of the Namibian side results in more watercourses and pools and hence more potential for tenure over these distinct water bodies. Regardless, concerns about access to mulapos must be seen in the context of overall low levels of concerns about access among Namibians.

A similar situation was reported in case studies by Hitchcock (1995) and Olomola (1998) regarding access rules for river fishing in Botswana and Nigeria, respectively. While outsiders were expected to ask the ostensible 'owner' of a body of water for permission to fish, in most cases permission was given and it was only in extreme situations, such as drought, that exclusion might occur. The broad and varying range of prohibited fishing measures reported by fishers is likely due to the fact that the survey spanned two provinces and wards with traditional authorities that have potential roles in fisheries management.

These overlapping institutions have led to a mosaic of different fishing management policies and enforcement measures. This finding is also reflected in a related study of fisheries regulations and the level of traditional authorities, where specific prohibited measures varied between the boundaries of each traditional authority along a 100 km stretch (Purvis et al. 2003). This situation is typical of transboundary resources, with multiple users responsible to different authorities, with different rules, capabilities and means of enforcement (Jones and Chonguica 2001). The significant difference between the number of Zambians and Namibians believing that no methods are prohibited is likely accounted for by the fact that Zambia has standardized rules that, while not evenly enforced, are likely known by fishers.

The difference in the apparent role of government and traditional authorities in each country is also influenced by the priority each country's government gives to inland fisheries. In Zambia, fisheries are part of a cabinet-level Ministry, with decentralized offices throughout the country. Thus, the national government has played a role in local fisheries management to date. In Namibia, fisheries in general have a more influential profile as a government responsibility; however, most emphasis is given to marine, rather than freshwater, fisheries. At the time of research, there was no inland fisheries legislation and limited extension services. Moreover, Namibian fisheries administration is highly centralized.

At the time of the survey, the closest inland fisheries office to the Caprivi was located over 1,000 km to the south. Thus, while government is ostensibly responsible for management, the present institutional and policy vacuum means that fisheries management is limited, explaining the perception amongst most Namibian fishers that no methods are prohibited. Illegal fishing was reported to occur, with Zambians being identified as the most common culprits by fishers on both sides of the river.

This can be explained in a number of ways. The absolute numbers of fishers on either side also make it more likely that Zambian fishers are illegally fishing, at least visibly so, or that conflict between fishers involve Zambians. The geography makes the Namibian side more desirable, so fishing without permission or out of territory is more likely.

Diversity of ethnic groups in Zambia means that the outsider/insider divide is not just between Zambians/Namibians—Zambians identify other Zambians as part of the problem.

The paramount Chief and other representatives of the traditional authority in Mwandi (Zambia) area have cited a growing trend of 'outsiders' moving from other areas of Zambia and establishing themselves along the river.

Given the prevalence of fishing as a livelihood in households with access to fishing throughout the study area (Abbott 2005) and the perception of illegal fishing occurring, it is not surprising that an albeit low number of fishers report fisheries-related conflicts. Zambians are cited as the most frequent group with whom conflicts occur by both sides.

However, the causes of conflicts are different on either side of the border. Namibians identify access and methods as being equally important sources of conflict but hardly mention the issue of too many nets, whereas Zambian conflicts are spread equally amongst the three categories. One potential explanation for this difference is that due to the floodplain's topography, tenure-related water bodies (such as streams and mulapos) occur more frequently on the Namibian side; hence asserting tenure over these areas may be a greater issue for Namibians.

Only three-quarters of Zambian fishers and less than half of Namibian fishers perceived a need for fisheries management. The majority of fishers that did see a need for management supported management by government over traditional authorities. The stated support for government may be biased by the fact that the frame survey itself was led and conducted in part by government staff. The higher level of support amongst Namibians for management by traditional authorities may be due to the history of devolution of responsibility for fisheries to the local level, which had only lapsed into ambiguity in 1990. By contrast, traditional authorities on the Zambian side of the river had experienced increased curtailment of their powers, both related to fisheries and in general, upon independence in 1964 (Bell-Cross 1974; Flint 2003). Limited backing for management by both government and traditional authorities could be due to the novelty or inherent ambiguity of the concept of co-management itself (e.g., Agrawal and Gibson 1999), especially in areas with overlapping national and traditional administrations and borders.

Perceptions about the rationale of regulations gave a promising result, as most Zambians and Namibians identified with the concept of conserving fish stocks, rather than excluding others as the goal. This is cause for optimism that regulations will be used in a productive way to overcome issues of access and method cited, despite the current environment of conflict.

What implications do these results have for emerging hybrid CBNRM and TBNRM efforts in the area? We address four important aspects of this question, namely, the growth and asymmetry of population and fishing effort in the area, similarities in fishing assets and activity, sources of conflict (related to fishing and otherwise) and management capacity.

The asymmetry in settlement patterns has two potential consequences for CBNRM and TBNRM. The number of settlements, high population, and ethnic diversity on the Zambian side of the river will make CBNRM more complex, as there are more stakeholders to consider and more 'communities' to account for. Community-based management in Namibia may be less complicated and Namibians were more supportive of management by traditional authorities. For TBNRM, Zambian fishers may claim larger representation and access to more fishing, particularly the productive fishing grounds in the Namibian portion of the floodplain, as Zambians are both more numerous and their settlements are more permanent. Namibian fishers would undoubtedly perceive their minority status under TBRNM with some concern.

The significance of potentially higher levels of fishing and smaller mesh sizes to CBNRM and TBNRM is ambiguous. Highly variable fisheries such as floodplains appear to be relatively robust to high levels of exploitation compared to temperate fisheries (Jul-Larsen et al. 2003; Welcomme 1985), meaning that limited management intervention may be needed. Furthermore, the similarity in fishing assets and patterns of exploitation amongst fishers implies that what management intervention might take place could be relatively simple and uniform, a prospect appealing to proponents of both CBNRM and TBNRM.

Nevertheless, the asymmetry in topography means differences in tenure and management demands can lead to 'conflicts of assignment' (Ostrom et al. 1994), when fishery resources are not spread evenly, causing some fishing grounds to be more productive and hence more desirable, than others. In effect, conflicts may arise not due to resource scarcity, but rather resource distribution (e.g., Turner 2004, 871). Potential conflicts over access to fishing may be especially apparent since most of the floodplains and in turn, the most suitable fishing grounds, lie within Namibian territory.

Existing sources of fishery-related conflict reported in our results lend support to the significance of asymmetry in settlement. Most respondents identified Zambians as being the greatest source of conflict, a reasonable finding considering that Zambians are the most numerous. The ethnic diversity of Zambian settlements makes CBNRM and TBNRM more challenging. From a broader context, inhabitants of the area are likely to invoke a broad range of identities, ranging from family to region, influencing how claims to different resources are asserted. These fluid points of reference do not easily fit into the view of a repaired social fabric envisioned by some advocates of TBNRM.

Similarly, the differences in levels and degrees of management as well as in the capacity to manage inland fisheries challenge efforts to make management locally relevant and consistent under CBNRM/TBNRM. The Namibian government has a limited (but growing) presence in fisheries management in the region, both in terms of infrastructure and local legitimacy. The mixed views regarding the most appropriate institutions for managing fisheries pose similar potential difficulties.

The majority of fishers' support for government management in either country reflects experiences with specific governments, which differ considerably in the case of Namibia and Zambia. In Zambia, the national government is responsible for fisheries management and indeed assumed considerable responsibility from traditional authorities upon independence. By contrast, until 1990, Namibian government policy mandated responsibility for fisheries management to traditional authorities. Hence, support for the government by Zambian and Namibian fishers may carry different assumptions of the role of other institutions.

The rationales that fishers (as well as other stakeholders, including NGOs and states) use to support different institutions may arise not out of an institutions' potential to manage, but rather the degree to which its legitimacy is reinforced. For example, traditional authorities and NGO-TBNRM initiatives may both back community-level management, but for different reasons. Support of CBNRM legitimizes already existing traditional authority structures; whereas for NGOs, a focus on the local scale makes national boundaries seem ail the less relevant, backing a rationale for TBNRM. Fishers meanwhile may support government management because it has been so limited to date.

Current efforts to incorporate aspects of community and ecosystem in the management of transboundary resources in southern Africa have been critiqued on the basis that differences in inhabitants' characteristics, livelihoods and attitudes towards management are overlooked, or at best assumed to largely correspond with goals of regional collaboration. These issues become particularly important when there is a high potential for variation in livelihoods and management, such as when neighbouring states have contrasting socio-economic environments, or the livelihood itself is inherently variable, such as artisanal fisheries.

We have described an example of a shared fishery resource in a river that also acts as an international boundary. The low level of management reported throughout the area is characteristic of floodplain fisheries and implies that a hybrid arrangement could be made amenable to variable intensity of fishing and movement of fishers. Despite the apparent differences in the socio-economic and political environments of Namibia and Zambia, fishers on both sides of the river have similar assets and fishing behaviour. There was support amongst fishers surveyed on both sides of the river for management of the fisheries. Moreover, the most frequent rationale reported by fishers to justify management, namely, the conservation of fisheries, is encouraging as it does not immediately presume that one group has more of a claim over resources than another. The conflicts reported by Namibians are largely linked to requesting permission to fish in an area, rather than outright encroachment. This suggests that negotiation, rather than strict rules of tenure and access, may be a more suitable approach to manage the area's fisheries.

It is tempting to assume enough common ground and flexibility exists between users, livelihoods and institutions to make hybrid fisheries management an appealing option. Our findings do not necessarily contradict this view. However, we underline the constant risks of oversimplification inherent in CBRNM and TBNRM and how these conceptual flaws may be magnified in hybrid management of resources. Management of the fishery as an ecosystem addresses fluctuations in the availability of fish and fishing areas. However, the spatial asymmetry of fishers and fishing effort makes it likely that management at the community scale will focus on access and use of Namibian resources by Zambian fishers. The greatest challenges in attempting to blend ecosystem and local scales will not arise from how it affects the management of the fishery, but rather on the myriad of social and political identities that are reinforced, challenged or created by the process.

8

Managing Water: Appurtenancy Reconceptualized

Until recently, the eastern United States has been blessed with an abundance of water; unlike the arid West, shortages in the East have historically been "rare and short-lived." During the past few decades, however, water has increasingly become scarce, due to recurring droughts and burgeoning urban and suburban populations. A severe drought struck the East Coast in 2002, forcing residents in rural Maine to stop flushing toilets and washing dishes, significantly damaging crops in the Midwest, and leading New Jersey to ban lawn watering. But that drought was only the latest and harshest of several that has hit the region since the 1980s. Rapid population growth and commercial development in and around eastern cities have also contributed to shortages, in some cases leading to interstate disputes.

Scientists predict that erratic precipitation patterns will persist, producing recurring droughts in years to come; continued growth in demand will only exacerbate the consequences of weather shifts. For policymakers, water shortages present three central challenges. First and most obviously, scarcity reminds us of the limits of our natural resources, and forces us to consider conservation measures. Second, scarcity requires that we assess the efficiency of water allocation and usage. Third—especially in the context of water, a universally necessary resource—equity demands that basic needs be met, and that one group of users not be allowed to exclude another group from an essential resource. In recent decades, water shortages have compelled eastern lawmakers to face these three issues—conservation, efficiency, and equity—directly. The result has been growing dissatisfaction with the legal regime that had historically governed the distribution of water rights in the East: common law riparianism. Under riparianism, "an owner of land abutting a waterbody has the right to have the water continue to flow... subject to the equal rights of each owner to make strictly limited use of the water." Riparian landowners, also called riparians, "share the common benefits that arise from adjacency to defined bodies of water."

Although riparian law had governed surface water in the East from the time of the first Anglo-American settlers, it came under increasing criticism during the latter half of the twentieth century. From a conservation standpoint, riparianism failed to protect against excessive diversions by riparians, or to ensure minimum stream flows for the public. From an efficiency standpoint, riparian rights were inherently uncertain because they were correlatively defined, and thus could shift over time as neighbouring users and uses changed; uncertainty, in turn, inhibited investment and prevented the development of markets in transferable water rights. And from an equity standpoint, riparianism was accused of favoring private agrarian interests.

Moreover, because rights were determined by ex post litigation, riparianism—like any common law regime—lacked a procedural mechanism for reallocating resources to those in need during times of scarcity. In the hopes of addressing these doctrinal shortcomings and better dealing with water shortages, approximately half of the eastern states formerly governed by common law riparianism established water permit systems, also known as permitting systems. These regulated riparian statutes—often adopted in response to scarcity concerns —envision a rationalized, administrative approach to water rights allocation. In theory, regulated riparianism requires that all direct users obtain a limited-duration, renewable permit from a state administrative agency.

This presumably allows the agency to make comprehensive ex ante decisions about preferred uses and impacts on other users, rather than resolve single disputes in ex post litigation. Agencies also have authority to reallocate water during times of shortage. While these regulatory schemes are usually meant to be comprehensive, certain groups of users are often exempt from the permit requirement. In contrast with common law riparianism, use on nonriparian land is generally allowed under permitting systems. What has become apparent over the last few decades, however, is that regulated riparianism suffers from its own weaknesses and, more importantly, has failed to solve many of the problems associated with traditional riparian law. Because of the exemptions accorded large groups of users, regulated riparianism appears "piecemeal," incoherent, and protective of inefficient uses. Even in those cases where permits are required, agencies tend to favour existing or grandfathered uses, thus preventing reallocation from lower-to higher-value uses. In addition, because permits create rights that are temporary and often "use-specific," voluntary transfer markets have failed to develop. Agencies also often fail to institute conservation-oriented norms in their permitting decisions. Finally, agencies have yet to develop satisfactory methods for allocating rights during shortages. In short, permitting systems have not proved much more successful than their common law predecessor in addressing problems of conservation, efficiency, and equity.

It is thus time for eastern policymakers to reassess regulated riparianism. Indeed, some states are currently in the process of intensely reviewing their water allocation laws, while those that lack permitting schemes are likely to impose some mode of regulation in the near future. Most eastern water law scholars believe that the answer is more comprehensive management and greater centralization of control. Richard Ausness, for instance, calls for the abolition of exemptions for large-scale users, increased centralization and coordination of administration, and more detailed planning for shortages.

Joseph Dellapenna likewise characterizes less-than-comprehensive schemes as "an unseemly hodgepodge of requirements," and advocates the creation of "effective comprehensive planning mechanisms." But is a policy of greater centralization and increased managerial control necessarily the wisest approach? Are there alternative governance regimes that might achieve greater efficiency, higher levels of conservation, or a viable market in water rights? Drawing on the extensive common property literature, this Note argues that such alternatives are indeed available.

A wide body of scholarship in anthropology and institutional economics demonstrates that local users in a variety of contexts, from lobster fishing to irrigation, have been able to develop sustainable methods for managing scarce, common-pool resources, thereby overcoming a tragedy of the commons without the need for government regulation. This Note asserts that the organizational and institutional insights offered by common property scholarship are relevant to contemporary water management. The Note begins by suggesting that the common property model may not be such a remote concept in eastern water law after all. Indeed, close consideration of traditional riparian law reveals the outlines of the common property model embedded in the doctrine itself. Specifically, argues that the appurtenancy requirement within riparian law—restricting riparian rights to appurtenant owners and restricting usage to appurtenant land-bears distinct similarities to key features of common property regimes. Limiting rights to those owning riparian land functioned as a proxy for creating an exclusive community, while restricting usage to appurtenant land capped, in a rough way, total usage of the resource.

Drawing on the appurtenancy doctrine, as well as the common property literature, I urge greater attention to certain central principles of resource conservation and management—principles that the appurtenancy doctrine, properly understood, can be seen to support. I should be clear from the outset about the goals of such an interpretation of the appurtenancy requirement. Identifying similarities between the appurtenancy doctrine and common property regimes is meant to illuminate structural values that have proved durable in a variety of contexts. In particular, I emphasize the success of groups of users who are closest to a common-pool resource at managing against its scarcity. For critics of riparianism who may be skeptical of the wisdom of returning to the common law for inspiration, I would stress that it is appurtenancy's underlying principles—the principles of proximity to, and familiarity with, the resource at stake—and not merely the formal doctrinal rules that continue to bear relevance.

In fact, the rigorous requirements of the appurtenancy doctrine have been significantly modified in the last century and a half, and in many jurisdictions they no longer strictly govern. Part III examines case law reflecting this doctrinal shift and concludes that the decline of the appurtenancy requirement was the inevitable result of another key doctrine in riparian law: the reasonable use requirement, under which riparian rights are correlative and are limited only by the rights of other riparian proprietors. I argue that whereas the appurtenancy doctrine relied on a fundamental limiting approach to resources, the reasonable use doctrine, as interpreted and applied by the courts, relied instead on an inherently tolerant and expansionist ethic. Contrasting these two strands within riparianism serves several purposes: It provides a doctrinal explanation for the weakness of the appurtenancy requirement, it suggests a historical account of the law's failure to conserve, and it highlights a distinction—important for resource-management purposes—between rules that limit usage and those that balance users' interests.

It returns to the context of regulated riparianism, and argues that the appurtenancy and reasonable use doctrines have important implications for ongoing management of surface water in the East. I first note that existing regulatory schemes tend to replicate reasonable use doctrine and conclude that, in their current form, they are unlikely to effectively resolve scarcity problems. I then explore potential alternatives to comprehensive regulation that might better reflect appurtenancy's respect for those closest to a resource and their ability to conserve. While recognizing that small-scale community management may be unrealistic in the eastern water context, and a return to appurtenancy proper undesirable, I nevertheless urge consideration of institutional models combining greater local control with broader government oversight. Such "nested enterprises" can offer the advantages of both the commons and the state, while potentially mitigating their respective weaknesses; by incorporating appurtenancy into the modern regulatory context, they may thus provide the best opportunity yet to improve conservation, achieve greater efficiency, and ensure equity in eastern water usage.

The Appurtenancy Requirement as Common Property Regime

At its core, riparian water law is about the entitlement of those bordering on a watercourse to certain use rights. The word "riparian" derives from the Latin ripa, which means the bank of a stream. Under the appurtenancy doctrine, a component of riparianism, only those who own property abutting the relevant body of water are entitled to access and use the water; in addition, water can only be used on adjacent parcels of land. This dual requirement has been subjected to substantial criticism, characterized as outdated and "anachronistic," accused of burdening urban and industrial expansion, blamed for "the development of uneconomical 'bowling-alley' parcels of land," and denounced for impeding the creation of marketable water rights and the increase in economic efficiency that many claim would follow. Indeed, overcoming the appurtenancy doctrine has been cited as "one of the principal motives [behind] the enactment of a regulated riparian statute." This Part reconsiders the appurtenancy doctrine in some detail and asserts that its underlying principles may prove more useful for resource management than its critics typically suggest.

The idea that the restrictions imposed by riparian law may have functioned as a common property regime has not gone unnoticed. Carol Rose, in arguing that common property regimes are more common than is generally recognized, briefly notes that "riparian law in the nineteenth century effectively turned river-bank landowners into participants in common property regimes... from which outsiders were excluded." Building on Rose's intuition, this Part fleshes out in more detail the various ways in which riparian law created, or at least attempted to create, a common property scheme, and compares this scheme to common property regimes in a variety of contexts. My goal here is to illuminate key principles for governing scarce resources.

The Ownership Requirement

Access Restrictions in Common Property Regimes : Under traditional riparian law, only those who owned property bordering a watercourse were entitled to riparian rights. Conceptually, this ownership restriction limited access to water to a community of easily identifiable users. It is what Henry Smith would characterize as an exclusion rule, and it has proved central to conservation and effective management of common-pool resources in a variety of contexts. Exclusivity facilitates the development of social norms that favour conservation of scarce resources. James Acheson, in his study of Maine lobstermen, describes harbor gangs as exclusive groups in which membership is usually necessary in order to gain access to the resource. Indeed, some gangs require land ownership in order to gain entry, thereby paralleling riparianism's appurtenancy requirement. Because groups are limited in size, members are more likely to form "long-term, multistrand ties with one another," which in turn generate norms about "'proper' behaviour." In the fishing context, proper behaviour means eschewing overfishing: "There is a strong feeling that all members should have equal access to the resource.... Too much success, or success earned the wrong way... can generate envy, hostility, and gossip." Lobstering is recognized as a zero-sum game, and those who play too aggressively are disliked. Indeed, it is noteworthy that the most exclusive gangs, which place the greatest emphasis on longstanding kinship and community ties, are the ones that have "voluntarily agreed to limit the number of traps they fish."

Studies of Japanese mountain villages, Spanish huerta irrigation communities, Cree Indian fisheries, and Kwakiutl Indian potlatching practices likewise emphasize the relationship between exclusivity and sustainable resource usage. These communities developed detailed harvesting rules or fishing practices that prevented exhaustion of the resource at stake. As Bonnie McCay and James Acheson put it, territoriality—or "restricting access to a defined body of people who inhabit or lay claim to a particular territory"—provides "the basis for the development of more restrictive common property institutions.... If we can keep others out, it makes sense for us to do something about our own behaviour." In these communities, to which access is often strictly guarded, centuries-long traditions and "trial and error" experience breed conservationist norms.

Limiting group size also facilitates the monitoring and enforcement of usage rules. Small, enclosed communities benefit from "rich opportunities for mutual monitoring," which reduce the cost of finding violators. Both Acheson's lobster gangs and Spanish huerta irrigation systems display these traits. Moreover, because community members are repeat players, the threat of social sanctions and disrepute are often enough to produce adherence to usage rules. Fikret Berkes notes that among Cree Indians, "[e]ffective social mechanisms ensure adherence to rules which exist by virtue of mutual consent.... People who violate these rules suffer... social disgrace." Similarly, social prestige drives Kwakiutl Indians to engage in lavish potlatching practices, which D. Bruce Johnsen sees as a way to ensure territorial exclusion.

Finally, exclusivity allows community members to maintain a sense of control and autonomy. As one commentator has pointed out, "[P]roducers [in a common property situation] seem to oppose and flagrantly to violate any outside restriction on their practices, even when they would almost certainly regulate themselves if left in control." Barton Thompson's study of water institutions in the western United States illustrates the same point. In these institutions, where outsiders are regarded somewhat suspiciously, local users have been able to implement customized rules to conserve scarce water supplies much more successfully than has the state.

The Ownership Restriction in the Common Law : Common property regimes thus succeed in conserving resources in part by excluding others. Exclusivity maintains user control, provides opportunities for extensive monitoring, and facilitates the development of restrictive use norms by those closest to the resource. As Anthony Scott and Georgina Coustalin have noted, restrictions on transferability of riparian water rights likewise seek to protect "the exclusivity of the right in the stream-wide group of users." I therefore turn to cases from the nineteenth and early twentieth centuries, which—while perhaps not always strictly adhering to the appurtenancy requirement demonstrate its inherently exclusionary tendencies.

For instance, in cases dealing with grants of water rights to nonriparians, early courts held that such grantees were not entitled to full riparian rights. In Gould v. Eaton, for example, the defendants, who were nonriparians, had contracted for the right to place a tunnel on the grantor's riparian land in order to divert water. The California Supreme Court affirmed an injunction prohibiting the diversion and explained the extent of the defendants' entitlement under such a grant: "[It] gave [defendants] no right in the land which is adjacent to the stream, and they took by the contract no riparian rights in the waters of the stream...." While the defendants might have had a contractual right that was good as against the grantor, they had no rights as against other riparian proprietors. They were, in effect, excluded from the riparian community. Other courts have come to the same conclusion.

In one such case, Roberts v. Martin, the plaintiff was a gristmill operator, and the defendants owned nonriparian land near the plaintiff's parcel. In addition, the defendants purchased from an upstream riparian owner the rights to one of the springs on his land. The court held that their diversion of water violated the plaintiff's riparian rights. The court noted that the doctrine of reasonable use as between riparian proprietors did not include "'the case of an interference by a stranger, who, by any means, or for any cause, diminishes the flow of the water; for this also is wholly wrongful, and no question of the reasonableness of his action in causing the diminution can possibly arise.'" The court in essence announced a per se rule against water use by nonriparians, or "strangers," under which no such use could ever be found reasonable.

Perhaps more ambiguously, in Williams v. Wadsworth the Connecticut Supreme Court recognized the rights of a nonriparian owner based on his longstanding usage and acceptance within the riparian community. The plaintiff, a nonriparian, had purchased in 1851 from a riparian landowner "the right to build and maintain upon the land a dam," from which he diverted water via a pipe to his house. He used the water for domestic purposes, and also sold some of it to other town residents. The defendant, who owned a one-acre upstream parcel, began in 1881—thirty years after plaintiff had begun his usage—to divert water via "an acqueduct of his own... to his residence and premises," no part of which was riparian. The plaintiff's brief noted that after considerable expense, the defendant succeeded in supplying himself with water, which he used for domestic and culinary purposes, for watering his horses and cattle, for supplying a fountain or fountains, and for sprinkling his lawn and the street in front of his dwelling-house. A considerable quantity of the water thus diverted has run to waste, and none of it has been restored to the stream.

The court upheld the claim of the nonriparian plaintiff as against the riparian defendant, finding that he had acquired a prescriptive right through adverse use for the required statutory period. Given the uncertainties that often surround prescriptive rights in riparian law and the narrow scope courts generally give to such rights, it seems likely that the court considered other factors in finding the right. Indeed, the court took special notice of the fact that the plaintiff had secured from "all lower proprietors" a release by deed from any claims for his diversion.

Moreover, at the time that the dispute arose, the plaintiff had been using the water for approximately thirty years, while the defendant had been diverting for less than three years. The court thus seemed to find that the nonriparian plaintiff's usage—because longstanding and because explicitly accepted by the other riparian proprietors—was riparian, while the riparian defendant's diversion was not. Although the case might be read as an equitable exception to the ownership requirement, it can also be interpreted as confirming the community principle behind the appurtenancy requirement.

The release from liability that the plaintiff had sought and received from the other riparian proprietors, and his long-accepted diversion of the water, demonstrated that his usage had survived the community's scrutiny. In other words, his nonriparian status was outweighed by his proven entry into the web of interrelationships and monitoring that characterizes a common property regime. Finally, in a number of cases, courts granted plaintiffs injunctions against nonriparian diversions in part out of a concern that such nonriparian usage might ripen into a prescriptive right. The Roberts court, for example, noted that "[t]he unlawful act of defendants will, in time, ripen into an adverse right if permitted to continue. Equity affords complete and speedy redress in the premises." Such cases may have reflected the fear that a nonriparian user might forcibly gain access to the resource without properly entering the riparian community, again confirming the exclusivity principle behind appurtenancy.

The ownership component of the appurtenancy doctrine thus arguably gestures toward a key feature of common property management—limiting access to a group of users defined by their proximity to the resource at stake. A wide range of scholarship illuminates the ways in which such a restriction can assist in the conservation of a common-pool resource, and early case law reveals such dynamics guiding courts in their decisions. While exclusivity in itself is not necessarily admirable, other ideas central to appurtenancy—the ability of those close to the resource to develop sustainable practices and to monitor each other's usage, and the importance of user control over resource management—deserve attention in the establishment of any regime governing a scarce resource.

The use Restriction

The other component of the appurtenancy doctrine is its use restriction, under which water may only be used upon riparian land. In this Section, I argue that appurtenancy and its sister doctrines operated more or less as a cap on total usage. I examine a number of cases demonstrating courts' sense that riparian land within the same watershed served as a rough proxy for the amount that could safely be withdrawn from the stream. I then consider usage restrictions in a range of common property settings to emphasize the importance of setting limits on withdrawal, as well as the competence of those most familiar with a resource in determining the appropriate level of withdrawal.

The Use Restriction in the Common Law : The case of Williams v. Wadsworth illustrates the limitation principle at work. The plaintiff in Williams, though a nonriparian, had been diverting water, with the permission of all lower riparians, for approximately thirty years. When the defendant, a riparian, began diverting water to a nonriparian parcel, the plaintiff sued and was granted an injunction. The court's discussion of the extent of the defendant's riparian rights is telling:

> Being a riparian owner the defendant has the right to consume water upon riparian premises.... But this use is to be confined to riparian land. This limitation... stands upon the necessity for a restraining rule in order to secure something for all, and upon the presumption that the brook will supply the absolute needs of as large an area as is usually held in riparian ownership.

The court then warned that "[i]f land not riparian may draw to itself, equally with land riparian... then land not riparian may take precedence of land riparian and deprive it of water for either man or beast." In short, the court's decision seems to rely on a rough estimate that the amount of water needed to supply riparian parcels generally approximates the amount of water that may safely be withdrawn without depleting the stream. Some readers will likely question how effective the appurtenancy requirement could be at confining total usage, given that a riparian landowner could in theory purchase larger and larger parcels of land and, under the unity-of-title rule that applies in most riparian states, could thereby exercise her riparian rights on an increasing area of land. In response, I would emphasize that the court's opinion in Williams is offered not to prove the conceptual limits of the use restriction, but rather to illustrate the underlying principles at work in the case law—conserving against scarcity and relying on the judgment of user-owners as to parcel size. In addition, and perhaps more importantly, the appurtenancy doctrine did not function alone, but rather operated in conjunction with other limitations on the total area of land that could benefit from riparian water rights.

First, many courts applied a "watershed limitation" in addition to the appurtenancy requirement; the conjunction of the two rules reflects a desire to limit withdrawal from the stream. In Anaheim Union Water Co. v. Fuller, the court held that "[l]and which is not within the watershed of the river is not riparian thereto, and is not entitled... to the use or benefit of the water from the river, although it may be part of an entire tract which does extend to the river." Although a parcel might be contiguous to a river, only that portion of the parcel within the river's watershed was considered riparian. The court went on to explain the logic behind this doctrine:

> The principal reasons for the rule... are that, where the water is used on such land, it will... return to the stream, so far as it is not consumed, and that, as the rainfall on such land feeds the stream, the land is... entitled, so to speak, to the use of its waters.

Or as Richard Ausness explains, "The watershed limitation is based on the assumption that land beyond the watershed is outside the boundaries established by nature for riparian ownership...." In other words, the watershed requirement aimed to preserve the body of water in the stream by allowing it to be used only upon land that naturally returned such water to the stream. In this way, the appurtenancy and watershed rules worked together to reduce the effects of withdrawals. Some courts imposed even further limitations. For instance, the court in Town of Gordonsville v. Zinn limited riparian usage not only to land within the same watershed but also to land acquired in a single transaction. The plaintiff town owned a reservoir that supplied its residents with water. The defendant owned a very small parcel of riparian land—a twenty-five-foot strip—upstream from the reservoir, as well as a parcel downstream, upon which her house was located. Although the downstream parcel was also riparian, it had been acquired separately from the upstream parcel. The town sued the defendant when she began diverting water from her upstream strip to her downstream parcel, since water thus diverted bypassed the reservoir altogether. The court ultimately held for the plaintiff, enjoining the defendant from further diversion.

Putting aside for the moment the plaintiff's distribution of the water to nonriparian town residents, the court's opinion reveals a sense, similar to that displayed in the Williams opinion, that the size of riparian parcels will generally protect against depletion of the water source. Although the holding ultimately relied on the fact that the downstream receiving parcel was in a separate watershed from the upstream diverting parcel, and thus was nonriparian with respect to that diversion, the court nevertheless made sure to expound upon the single-ownership rule. Quoting from Farnham's treatise on water rights, the court took note of :

> *"the rule that all land must be regarded as riparian which is within the natural watershed of the stream, the title to which is in one owner and the boundaries of which have been established in accordance with the requirements of the conditions which will best serve the interests of individual landowners.... The watershed should certainly form a limit beyond which the riparian rights cannot be claimed; but there is a question whether or not that limit is not too wide.... Under this rule the boundary of riparian land is restricted to land... acquired by one transaction."*

Guarding against the possibility that land within a watershed, or separate riparian parcels held by a single owner, might "'recognize such a vast extent of territory as riparian land [so as to be] destructive,'" the court thus seemed to rely on the judgment of individual purchasers by using parcel size as a rough proxy for the area that could profitably be supplied by the stream in question.

Capping Mechanisms in Common Property Regimes : Taken together, these cases suggest that the appurtenancy requirement and related doctrines functioned as a rough cap on withdrawal, by relying both on naturally arising drainage patterns and the judgment of riparian owners. The appurtenancy and single-transaction requirements in particular exhibited the law's faith that riparian proprietors would purchase parcels of sustainable size. I conclude this Section by examining a number of analogous property regimes in which usage rules function as capping mechanisms, restricting the amount that may be withdrawn from a common-pool resource. This comparison places appurtenancy's reliance on riparian owners within the context of a range of mechanisms adopted by appropriators to voluntarily restrict their own resource usage.

Multiple scholars have identified voluntary limits on total usage as a key factor in the sustainability of common property regimes. In describing the common grazing areas under the English open field system, for example, Smith notes that not only was access to common grazing areas restricted "to a small closed class of appropriators," but further restrictions on usage also applied. In particular, "stints," or limits on the number of animals that could graze on the commons, essentially functioned as a capping mechanism, since "one sheep cannot eat much more even if it is allowed to graze longer." Stints, like the appurtenancy requirement and its sister doctrines, served as "proxies for harmful activities"; the eating capacity of sheep, like the size of riparian parcels, inherently limited total usage. The success of the open field system, both in increasing productivity and in overcoming the potential tragedy of the commons posed by communal grazing areas, has been well-documented. Stints, which proved extremely "effective and widespread" in preventing overuse, were a central component of this success.

Other examples of voluntary restrictions on usage abound. Acheson credits lobster gangs' self-imposed reductions in fishing efforts—in particular, voluntary limits on the number of traps, which necessarily limited total catch —for "both economic and biological benefits," including greater lobster density and higher lobstermen profits. Swiss villages have sustained a communal alpine economy since the thirteenth century by requiring that officials first "mark the trees ready to be harvested," which allows for "a careful assessment of the condition of the forest." In the extensive common lands regulated by Japanese villages for centuries, "detailed authority rules specif[ied] in various ways how much of each valued product a household could harvest from the commons and under what conditions." And, perhaps most pertinent to the appurtenancy context, more than one huerta irrigation system has limited water rights to certain areas of land; designations of some parcels as irrigated and some as dry have lasted for centuries.

Examining the use restriction of appurtenancy doctrine in light of the practices of common property regimes allows us to make two observations. First, from a resource standpoint, the restriction, while no doubt often inefficient and inequitable in confining water usage to adjacent land, may nevertheless have stemmed from a quite rational desire to limit withdrawal by users of a scarce resource. Appurtenancy and its sister doctrines may have tried to cap usage by drawing on the proxies of adjacency, watershed boundaries, and parcel size, just as medieval peasants relied on sheeps' digestive systems, and Maine lobstermen relied on the size of standard fishing traps. Second, as a structural matter, both appurtenancy and successful common property regimes seem to have relied on the intimate knowledge of proximate user-owners in determining the appropriate level of withdrawal. For common-pool resource management, both insights—about the need to cap withdrawal and the expertise of those closest to (and thus most familiar with) a resource in setting such caps—are essential.

Reasonable use and the Decline of the Appurtenancy Requirement

This Note began with the premise that, in light of continuing water shortages, eastern states need to reassess current water regulation and develop more sustainable resource-management practices. Resisting the preference exhibited by most commentators for comprehensive and centralized regulation, I instead turned to an unlikely place—the appurtenancy doctrine—and suggested that appurtenancy contained certain structural features that have proved successful in governing and conserving scarce resources in a variety of common property contexts. A complete account, however, must recognize that the success of riparianism in restricting access and usage has been, at best, quite limited.

Though modern treatises still define riparian rights as those "concerning the relation of the owner of the bank of a stream to various features of the stream," most commentators recognize that the appurtenancy doctrine—the feature that gives riparianism its very name —is today subject to numerous exceptions. "[E]very state," notes James Christman, "allows some means of transferring riparian rights apart from the land." Granted, the law regarding these transfers varies from jurisdiction to jurisdiction and remains far from clear: Some states recognize conveyances, though whether such conveyances are valid only against the grantor, or against all other riparians, is a matter of some dispute; a number of states recognize the right to lease riparian water rights; and a fraction of all states recognize prescriptive nonriparian rights.

Nevertheless, the general trend has been away from appurtenancy and toward increasing nonriparian use. Recent reformulations of the law likewise recognize the need for nonriparian rights. The Restatement (Second) of Torts does not consider "classification of [a use of water] as riparian or nonriparian" to be controlling, and the influence of this view is seen in court decisions—most notably Pyle v. Gilbert —that recognize nonriparian rights. And finally, the permitting systems and regulated riparian regimes that increasingly govern surface water in eastern states also recognize nonriparian rights as permissible. Why did the appurtenancy doctrine never really take hold, and what relevance does this have for modern resource management? In this Part, I offer two accounts. I examine cases in which courts made exceptions to the appurtenancy requirement; this doctrinal analysis leads me to conclude that the reasonable use doctrine that increasingly governed riparian water law during the nineteenth century inherently allowed for expansion of the riparian community and the acceptable set of riparian uses. Section B offers some broader historical reasons for the failure of the appurtenancy doctrine, emphasizing the cavalier approach to natural resources that characterized the United States during the era of industrialization. In the next Part, I draw on these conclusions and suggest that the tolerance for increased usage inherent in the reasonable use doctrine continues to have significance today.

Reasonable use as Doctrine

Under the reasonable use doctrine, "each owner of riparian land is permitted to use the water in a waterbody, regardless of the effect the use has on the natural flow, so long as each user does not transgress the equal right of other riparians to use the water." The rise of reasonable use doctrine during the nineteenth century, replacing the more stringent natural flow theory of riparian lights, has been the subject of considerable scrutiny. The argument here is that reasonable use, at least as framed by nineteenth-century courts, was a doctrine inherently tolerant of expanding uses of water; as such, it operated in tension with the appurtenancy doctrine, which I have characterized as inherently restrictive.

The classic statement of riparianism in American law is Justice Story's opinion on circuit in Tyler v. Wilkinson. In that case, the lower riparian plaintiffs claimed that upstream proprietors were entitled only to water that the plaintiffs did not need. The court, however, rejected this suggestion, holding that "[t]here may be, and there must be allowed of that, which is common to all, a reasonable use." While the opinion seemed to hold fast to natural flow theory in some respects, it also acknowledged what Morton Horwitz has called "the utilitarian criterion of valuable use [as] the ultimate source of legal rules." Specifically, Story rejected the idea that riparianism would allow for "no diminution whatsoever... for that would be to deny any valuable use." The Tyler case, while still nominally adhering to natural flow doctrine, thus opened the door to a balancing approach to riparian rights that would take into account "valuable use."

Further elaboration of the reasonable use doctrine has only added to its malleability. In the 1883 case of Red River Roller Mills v. Wright, for example, the Minnesota Supreme Court offered this definition of the factors to be taken into consideration in a reasonable use analysis:

In determining what is a reasonable use, regard must be had to the subject-matter of the use; the occasion and manner of its application; the object, extent, necessity, and duration of the use; the nature and size of the stream; the kind of business to which it is subservient; the importance and necessity of the use claimed by one party, and the extent of the injury to the other party; the state of improvement of the country in regard to mills and machinery, and the use of water as a propelling power; the general and established usages of the country in similar cases; and all the other and ever-varying circumstances of each particular case bearing upon the question of the fitness and propriety of the use of the water under consideration.

The reasonable use analysis thus became open-ended and case-specific. Such a utilitarian approach, concerned as it is with comparing social costs and benefits, is inherently open to expansion; it allows—indeed, it encourages—courts to take into account the economic value of the infringing user's activity. As Smith notes, such tort-based rules do not protect rights, but rather focus on "the efficient use of resources. Scarcity over resources is reduced to bilateral conflicts over use." The key considerations are "maximizing value" and "optimal resource allocation," rather than protection against global scarcity. A reasonable use regime is governed by considerations of "social values and needs" and "general rules of permissible and impermissible human conduct"; as such, it is inherently "fluid and socially responsive."

Finally, consider the focus on actual injury in the reasonable use inquiry. As Justice Story put it in Tyler, "The true test of the principle and extent of the use is, whether it is to the injury of the other proprietors or not." Courts often cite the doctrines of de minimis non curat lex and damnum absque injuria in rejecting plaintiffs' claims, finding damage nonexistent, or simply too small to justify a remedy. This rule, too, fits with an expansionist interpretation of reasonable use: Defendants will not be enjoined, even if their activities violate established rules of riparian use, unless plaintiffs can actually show injury.

Reasonable use's tolerance for expanded usage is perhaps most apparent in those cases allowing exceptions to the appurtenancy requirement. In these decisions, reasonable use essentially swallows up the appurtenancy doctrine. In Elliot v. Fitchburg Railroad, for instance, the defendants purchased a riparian parcel upstream from the plaintiff's land and diverted water to their depot to furnish locomotive engines. Although the diversion violated the appurtenancy rule, the court held the defendants "not liable" because the plaintiff had not "suffered actual perceptible damage." Reasonableness, it held, was "a question of degree," not the absolute rule appurtenancy would impose.

Similarly, in Jones v. Conn the court allowed the defendant to irrigate nonriparian land and power a gristmill over the complaint of the plaintiffs, who used the water to irrigate riparian parcels. In opposition to the theory of appurtenancy, the court noted that "whether a proposed use is... reasonable, does not depend so much upon the area of the land... or the place of use, as upon the effect it has upon... the other proprietors." Here, the "amount of water taken... was not sufficient to materially injure the plaintiffs," so the defendant could not be deemed a "wrongdoer."

Even more extreme is the Oklahoma Supreme Court's statement in Smith v. Stanolind Oil & Gas Co. The plaintiffs sought to enjoin Stanolind from obstructing water flow and taking the water for nonriparian purposes. The court emphasized that nonriparian use "standing alone, [is] no criterion of the reasonableness of the use.... [A] riparian owner has the right to make any use of water... possible, so long as he does not inflict substantial or material injury [on other riparians]." Doctrinally, one can therefore link the decline of the appurtenancy doctrine—or perhaps more accurately, the refusal of courts to insist upon its strict application—to the rise of reasonable use. The appurtenancy doctrine, which we might characterize as a resource-focused limiting rule, conflicted with the user-focused balancing rule of reasonable use. This account is thus somewhat in tension with Rose's description of the reasonable use doctrine:

> [R]iparian law in the nineteenth century effectively turned riverbank landowners into participants in common property regimes for particular rivers, from which outsiders were excluded (no interbasin water transfers) and insiders were expected to use the common resource "reasonably"—in effect, in a way that was compatible with equal correlative use by all the other riparian owners.

In other words, Rose suggests that both place-of-use restrictions and the reasonable use doctrine were a part of the common property regime. The view offered here, however, is that reasonable use actually worked in opposition to and defeated the restrictive common property elements of appurtenancy. While reasonable use initially may have governed a community of insiders, the doctrine's tendencies to treat uses and users "reasonably," to consider the social value of activities, and to decide in favour of efficiency and productivity, would inevitably lead to expansions of the community as well as the range of accepted uses.

While the Tyler test of injury initially looked only to the impact on, and competing claims of, other riparian proprietors, its principle of balancing and value maximization would eventually require consideration of the interests of outsiders. In short, reasonable use sooner or later tended to undermine appurtenancy. This account of reasonable use versus appurtenancy allows us to draw a conceptual distinction among different types of what Smith calls governance rules—a distinction not fully apparent from Smith's analysis, but important for resource-management purposes. Under Smith's taxonomy, governance rules involve the "specification of proper activities" for which a resource may be used.

Given this broad definition, both appurtenancy's use restriction and the reasonable use doctrine would qualify as governance rules, since both speak to appropriate usage: Appurtenancy specifies the parcels of land on which water may be used, while reasonable use (at least in theory) specifies the purposes for which it may be used. Smith's focus is on the relative information costs of different rules, and he thus differentiates between governance rules of greater and lesser specification. He does not, however, distinguish between rules that limit total usage and those that merely aim to balance conflicting users' interests. He sees governance, like exclusion, as an "alternative method of preventing dissipation," without recognizing that certain governance rules, such as the reasonable use doctrine, may actually facilitate increased usage of a resource. Smith notes that common property regimes often involve both exclusion rules and "rules of proper use among [group members]," which may "promote conservation and sustainability of common-pool resources."

But he does not acknowledge that some "proper use" rules emphasize limitation, while others may focus on value maximization—a distinction that is at the heart of riparian law's shift from appurtenancy to reasonable use. Smith does briefly address the transition from natural flow theory to the reasonable use doctrine, but asserts only that "exclusion tends to give way to governance as pressure on the resource has increased," without analysing the implications of reasonable use for a resource under pressure. His analysis thus misses the key difference among governance rules that this Note has emphasized in the two riparian regimes. Examining the limited success of the appurtenancy doctrine in relation to the reasonable use doctrine, and analysing the cases in which courts allowed the former to give way to the latter, one can see an important conceptual distinction—limiting versus balancing—among various regimes aiming to conserve scarce resources.

Reasonable use as Historical Development

The rise of reasonable use, however, is not an entirely satisfactory explanation of why appurtenancy failed to take hold. Indeed, doctrinal analysis simply begs the question: It explains what judges did, but not why they did it. In this Section, I briefly offer some suggestions as to why judges turned away from appurtenancy and constructed an ever-broadening notion of reasonable use.

Morton Horwitz's legal realist account holds that the rise of reasonable use and fall of natural flow was part of what he calls "the transformation in the conception of property" that occurred during the nineteenth century. The reasonable use standard was "an open-ended formula through which common law judges could implement their own conceptions of desirable social policy." What I, somewhat agnostically, as the value-maximizing tendency of reasonable use is viewed by Horwitz in more political terms, as "central... to the needs of industrial development... in the nineteenth century" and "essential to economic improvement." Rose summarizes Horwitz's explanation succinctly: "Legal developments [in water law] gave a kind of common law subsidy to capitalist developers, allowing them to inflict some injury upon weaker and less enterprising folk without having to pay for the consequences." Theodore Steinberg, an environmental historian, agrees: "The drive to promote industrial growth had, in the process, reduced water to an instrument in the service of economic change. Water by the early years of the nineteenth century was well on its way to becoming a simple utility, an abstraction employed to suit economic ambitions.... a commodity."

Other, less political explanations are available, and they complement Horwitz's account. As riparianism moved beyond New England, into southern and western parts of the country, it encountered regions where water was used less for power—an instream use that would not have found the appurtenancy doctrine unduly restrictive —and more for consumptive uses such as irrigation and farming. In these areas, a more tolerant attitude toward withdrawals by nonriparian owners for use on nonriparian land undoubtedly seemed more attractive. Moreover, once steam became a major source of energy, and the mill economy of the Northeast declined, "courts began to appreciate the fact that it would be efficient to increase the consumption of many streams and they began to develop a law of consumption through the reasonable use theory." In short, both economic and geographic expansion of the nation likely gave impetus to doctrinal change.

Finally, the need to expand the acceptable uses of water went unchecked by any sense of scarcity. Quite simply, the humid eastern states did not face the same shortages as western states. Courts lacked a pressing reason to adhere to the logic of conservation and restriction embedded within the appurtenancy doctrine. With water plentiful and the economy expanding, the dominant ethic of industrialization demanded the fullest use of natural resources. In such a climate, courts found little need to insist on ownership or use restrictions, and plenty of reasons to tolerate expanded access and withdrawal.

This Part has offered two explanations for the weakness of the appurtenancy requirement in riparian law. As a doctrinal matter, appurtenancy and its restrictive attitude conflicted with, and succumbed to, the more tolerant balancing approach of reasonable use. As a historical matter, appurtenancy's values of limitation and conservation did not fare well under the pro-development, efficiency-maximizing policies of the nineteenth and early twentieth centuries, nor did they seem necessary in light of the seeming abundance of water. This comparison of the reasonable use and appurtenancy doctrines proves helpful as we return, in the next Part, to regulated riparianism.

The Continuing Relevance of Appurtenancy's Principles

Having considered in some depth the two basic doctrines of riparianism—appurtenancy and reasonable use—as well as the features of common property regimes that have enabled users to sustain and conserve scarce resources, we are now able to draw several conclusions to aid us in our evaluation of regulated riparianism: First, restricting access to a resource to its closest users can facilitate the development and enforcement of conservation norms. Second, because users are generally resistant to external oversight, placing management authority in their hands will often increase the likelihood that they will adopt effective resource usage practices. Third, because users will generally have greater familiarity with the resource and its changing dynamics than government actors, they are often better able to determine appropriate levels of usage and to limit their withdrawals accordingly. Finally, regimes that rely on a value-maximizing, balancing approach are less likely to protect scarce resources against growing demand than regimes that take a more restrictive, limiting approach. With these observations in mind, I offer a critique of current water law regimes in Section A. While a full-scale policy proposal is beyond the scope of this Note, I do suggest, in Section B, a number of ways in which the lessons of appurtenancy and common property management may be applied to water law today.

Reasonable use Redux

Begin by drawing attention to the survival of the reasonable use doctrine under current regulatory schemes. Most statutes list a set of factors for agencies to use in deciding whether or not to grant a permit; such lists closely resemble the open-ended inquiries conducted under the reasonable use doctrine in cases such as Red River Roller Mills. The Regulated Riparian Model Water Code, for example, provides a list of nine factors to be taken into account by the permitting agency, including "the economic and social importance of the proposed water use and other existing or planned water uses sharing the water source."

Such lists "can only reinforce the sense that the substantive basis for decision remains much the same as in traditional common-law litigation under the reasonable use theory of riparian rights." While advocates of riparian statutes stress that such decisions are made ex ante, rather than ex post, I would question the extent to which moving the reasonableness inquiry from an ex post judicial decision to an ex ante administrative determination can counteract the tendency of such balancing tests to permit increased usage, instead of protecting scarce resources. As one scholar comments, "[I]n permit-based jurisdictions, allocative standards make little effort to distance themselves from their riparian ancestors."

Supporters of regulated riparianism may note that every permitting statute requires the administering agency to develop long-term plans to protect the source for future needs, and that almost all permitting statutes also "require the agency to define or protect minimum flows." But the existence of such statutory mandates does not ensure protection of water resources. Long-term plans, for instance, are often developed according to a list that, again, echoes the reasonable use inquiry; they are thus unlikely to result in greater conservation. Furthermore, statutory mandates to protect minimum flows have proved, in practice, difficult to administer, given "traditional water management" values and the "considerable discretion" granted administering agencies. In Arkansas, for example, where a regulated riparian statute provides for the protection of minimum stream flows, "[t]he process of setting minimum flows has proven highly political, [resulting in] considerable modification (and reduction) of the proposed original levels of protection." A series of recent agency and court decisions in Vermont has severely limited the ability of the state's permitting agencies to use the public trust doctrine to protect minimum stream flows. Lack of strong enforcement of minimum-flow protections, combined with the persistence of reasonable use inquiries in permitting decisions and long-term planning, may thus indicate that regulated riparianism will fare no better at conserving water supply than its common law predecessor.

Local-global Possibilities

Rather than press for greater centralization and control while adhering to reasonable use inquiries and multifactor tests, regulators should take a step back and consider potential alternatives to the comprehensive management model. Indeed, common property scholars have extensively documented the disastrous consequences—in both ecological and human terms—that can result from wrongheaded state intervention. In light of the analysis offered in this Note, I would urge serious consideration of regulatory responses that attend to scarcity in a way that incorporates the core principles of appurtenancy. In other words, while regulated riparianism has until now included primarily the reasonable use strand from common law riparianism, I would instead emphasize the appurtenancy strand.

Some readers will no doubt question the feasibility of returning to the strictures of the appurtenancy doctrine. Indeed, confining water usage to adjacent bodies of land, for the benefit of the owners of those parcels, seems neither a realistic nor a desirable goal for contemporary water management. I would suggest, however, that appurtenancy need not be conceived of solely as a physical restriction. Instead, I propose here an understanding of appurtenancy as a structural, organizational concept for resource management. It is this reconceptualized notion of appurtenancy that may prove valuable for policymakers today.

Reconceptualizing Appurtenancy

The traditional understanding of appurtenancy focused on physical proximity: Only owners whose property abutted the water had rights to it, and only their adjacent land could benefit. But as the cases reveal, and as comparisons to common property scholarship highlight, the doctrine was not solely about physical proximity, but also contained certain structural principles of resource management. Limiting riparian rights to adjacent owners excluded "strangers" and thus defined a group of users who could enforce conservationist norms of behaviour. Restricting usage to appurtenant parcels demonstrated a reliance on those most familiar with the resource to determine appropriate levels of usage. In other words, appurtenancy doctrine expressed a commitment to vesting resource management in those closest to, and sharing an interest in, the waterbody. In the eighteenth and nineteenth centuries, this structural commitment coincided with physical proximity; adjacency functioned as a proxy for defining the most knowledgeable community of users. But in a twenty-first-century world in which water consumption needs are vastly different, we should perhaps de-emphasize the physical requirements associated with appurtenancy and focus instead on its delegation of governance rights to those most intimate with the resource, whether physically proximate or not.

Under this definition, appurtenant users would no longer necessarily be those with property bordering a given resource, but would rather be defined as those within a given region using a resource in a similar way. Membership in a community of users would be conceived of not in terms of adjacent land ownership, but rather in terms of the particular use—whether irrigation, recreation, or municipal water supply. Appurtenancy, on this understanding, would vest management authority not in those physically proximate, but in those functionally proximate—both to the resource and to a like group of users.

Putting appurtenancy to work in a contemporary water regime would thus mean recognizing the competence of communities of users in effectively governing scarce resources. At first glance, the workability of such a governance scheme in the eastern water context might seem dubious. After all, the examples of community management generally involve close-knit groups that are able to exclude outsiders because of their small size, and whose ability to conserve resources depend largely on longstanding social ties. While such community-based regimes might work in the context of small streams or rivers, they seem implausible when it comes to larger watercourses that may span not only great distances, but also multiple jurisdictions. Moreover, unlike the irrigators in huerta communities or the farmers in medieval England, the heterogeneity of water users in eastern states makes traditional collective action seem unlikely to succeed. And while common property management has proved successful at addressing conservation concerns, its exclusionary nature and resistance to outside interests might not sit well with those worried about efficiency and equity.

In response to such concerns, I would first suggest that appurtenant groups need not be limited to homogeneous communities of individual users; users drawing on the same body of water for similar purposes could be considered appurtenant, even if they were institutional entities. Second, I would emphasize that small-scale community-based management and comprehensive government regulation are not the only options. We may do better if we avoid "the familiar trichotomy of private, communal, and state ownership," and instead pay more attention to the possibilities of localstate co-management. As Oran Young suggests, rather than "focusing exclusively on finding the right level or scale at which to address specific problems.... the key to success [may instead lie] in allocating specific tasks to the appropriate level of social organization and then taking steps to ensure that cross-scale interactions produce complementary rather than conflicting actions." Indeed, much recent work in common property scholarship and institutional economics more broadly has been directed toward the possibilities of local-state co-management. Thus, just as the Note drew on the first wave of common property scholarship to illuminate our understanding of the appurtenancy requirement, I now turn to more recent work in the area to inform our regulatory policies going forward. Specifically, I consider what Elinor Ostrom has called "nested enterprises," in which "[a]ppropriation, provision, monitoring, enforcement, conflict resolution, and governance activities are organized in multiple layers." Under such a model, respect for appurtenant communities could be embedded within a larger-scale government system.

While nested enterprises may sound intriguing in theory, readers may wonder whether such efforts work in practice, and how they would work in the context of eastern water law. In the remainder of this Note, I therefore describe successful examples of cross-scale management. I also suggest reasons to believe that such regimes may provide the advantages of both the commons and the state, while mitigating, rather than exacerbating, their respective weaknesses. I conclude by arguing that multilevel governance may be particularly well-suited to the dilemmas facing eastern water regulators. In particular, nested regimes that embed appurtenant groups within an administrative scheme may more effectively address the three major concerns conservation, efficiency, and equity—that surround allocation of our water supply.

Nested Enterprises: Some Models

Numerous instances of successful nested enterprises exist. This Subsection explores four models and considers in each instance how norms of usage and conservation developed by appurtenant groups can be effectuated within a larger government system. These examples highlight the ways in which nested enterprises may yield the advantages of both the commons and the state, while mitigating the weaknesses of each. One nested enterprise model, which we might call the reverse command-and-control model, utilizes state authority, expertise, and administrative resources to enforce usage roles developed by appurtenant groups to protect against scarcity. For example, the Maine lobster gangs have, in recent years, formed links with state government, which in turn aids them in the enforcement of their voluntary restrictions on fishing.

Acheson notes that the inhabitants of Monhegan Island have "persuaded the legislature to pass a law forbidding fishing in Monhegan waters from June 25 to January 1," thereby making official a customary closed season that the islanders have observed since the early 1900s. Fishers from Swan Island likewise convinced Maine's commissioner of marine resources to enforce a self-imposed limit on the number of traps surrounding the island, A similar example comes from the Aleut community in Alaska, where the community controls access to the territories, and thus can control the level of fishing, but has also "made the state limit fishing effort in adjacent territories where the local stocks might be intercepted by outsiders."

Yet another example can be found overseas, in the state of Kumaon in India, where centralized rules create "a framework for the management of forests rather than a defining straitjacket," and at the same time establish "a domain of relatively autonomous action and rule making in which local residents and their representatives can operate." Specifically, local authorities "make rules and enforce them," "facilitate some kinds of actions... and restrict others," meet frequently to modify rules, and "create monitoring and sanctioning mechanisms." The state government supplements these efforts by supervising recordkeeping, coordinating harvesting, providing technical assistance, and aiding in rule enforcement. As these examples demonstrate, reverse command and control takes advantage of the greater skill of local users in developing rules of conservation, and places those rules under the enforcement authority of government.

Another nested governance model that has gained attention recently combines market-based tradable allowances with community management. Under tradable allowance schemes, governmental regulators in effect place an upper limit or cap on the total quantity of a given resource that is to be available for use.... The regulators then divide the capped total into individual allowances. Henceforth they require all resource users to purchase or trade for whatever allowances they use.

In a nested enterprise model, quotas are allocated to appurtenant communities, rather than individuals or individual entities. In Alaska, for example, some of the walleye pollock quota was allocated to native communities; similarly, New Zealand transferred forty percent of its fishery quota to the Maori. Such schemes incorporate groups of users-proximate both to each other and to the resource—into a larger regulatory market, thereby protecting community interests while maintaining the advantages of a centralized decisionmaker for setting and managing overall goals. In these regimes, as compared to the reverse command-and-control model, the state does not undertake enforcement of specific usage restrictions developed by the user community. Instead, it grants an appurtenant group of users a block of rights that is presumably large enough to allow the group to maintain a sustainable level of withdrawal, leaving the group to implement and enforce its own conservation norms. At the same time, the tradable allowance scheme permits user groups to operate within a broader market of uses and users. The existence of a wider market in standardized rights can facilitate transfers from lower-to higher-value uses, thereby increasing the efficiency of resource usage.

The creation of standardized rights can also facilitate coordination across different appurtenant groups—functioning as a kind of common language and in some instances leading to larger-scale conservation efforts. A third model—what we might call the horizontal governance model—comes from voluntary associations of groundwater users in Southern California. These nested enterprises should be particularly persuasive not only because they involve water as opposed to fisheries or forest management, but also because they demonstrate the viability of such a model among institutional, as opposed to individual, users. The parties involved here are not members of an Indian village, or an Aleut fishing community, but include municipalities, utility companies, and private mutual corporations.

Elinor Ostrom documents the rise of "a set of institutions [established by water producers] to manage a series of groundwater basins located beneath the Los Angeles metropolitan area." Faced with the prospect of saltwater incursions, dropping aquifers, and uncertain rights, appropriators began to engage in a "pumping race," a classic example of a tragedy of the commons. During the 1940s and 1950s, many believed that the race would result in the destruction of the basins, with dire consequences for the water supply in southern California. But instead of succumbing to the tragedy, water producers entered into a series of negotiations, through which they were able to form institutional structures that combined public and private governance, to set restrictions on withdrawals, and to institute conservation measures. Nested within the public arena, private actors were able "to impose constraints on themselves."

A few features of these settlements are particularly noteworthy. First, the disputes arose among functionally appurtenant users—claimants with shared interests in the same body of water. Second, in coming up with a solution, it was important to the users that they maintain some measure of management authority; they did not want to turn decisions over to large-scale agencies, "for fear that they would lose control of the decisions being made and might end up worse off." In short, they exhibited the same desire for autonomy and self-governance as members of the common property regimes. Third, while users maintained decisionmaking control, they did so within a larger governmental regime; the institutional solution was neither deregulation nor centralization, but rather "polycentric" systems of shared private-public governance. While the first two models of nested governance are largely vertical and thus somewhat hierarchical, with the state at the top and user groups at the bottom, this model vests management authority in a more horizontal, or multipronged, manner. Fourth, the costs of monitoring and enforcement were shared by the users (two-thirds) and the state (one-third) And finally, the negotiations among parties led to a better clarification of rights, which in turn allowed a market to develop, leading to a transfer of rights to those using them at "a higher value."

The Ostrom study allows us to draw two important conclusions. First, appurtenant groups can be defined to include institutional users—here, the appurtenant groups were voluntary associations made up of both individual and institutional user-appropriators—who maY prove just as capable of developing and enforcing conservation norms as individuals within a small community. Specifically, institutional appropriators restricted their water usage and monitored each other's withdrawals, just as farmers in huerta villages restricted water usage to certain parcels of land and kept an eye on their neighbours' withdrawals. Second, a nested enterprise model in which management is "polycentric" rather than centralized, and in which appurtenant groups of organized users actively participate can produce an allocation of tradable rights that is efficient and stable, as well as conducive to conservation. As clarification of rights has proved difficult under regulated riparianism, and a market in permits has failed to materialize, the achievements of the entities Ostrom describes are particularly noteworthy. In short, the horizontal governance model seems to have had more success in resolving efficiency and conservation problems than have eastern regulatory systems.

A fourth model—what we might call the umbrella model—is illustrated by local water institutions in the western United States. Mutuals and water districts are the retail distributors of most domestic water and much agricultural water in western states. Mutuals are private nonprofit corporations whose customers are also shareholders; water districts, on the other hand, are governed by elected boards, like other local government units. "Both types of institutions engage in a broad set of activities, including obtaining and storing necessary water supplies, transporting the water to their service areas, and distributing it to their members." For our purposes, mutuals and districts may be analogized to appurtenant groups (though the water they consume may come from a distant reservoir or dam), as they allow groups of local users to access water jointly, thereby taking advantage of economies of scale, and to maintain control over allocation within the group.

These institutions have developed into multilayered regimes, thus exemplifying a more complex, cross-scale kind of nested enterprise than the models already discussed. Mutuals and water districts may obtain their water from a larger umbrella agency acting as a wholesaler, or they may sit within larger umbrella institutions themselves. All of these institutions ultimately sit under the general authority of the legislature. Several features of these institutions are familiar. For instance, whereas users have generally been resistant to state proposals for conservation, they have been receptive to conservation programs implemented by local institutions. These programs have been able to reconfigure groundwater rights; impose well-spacing rules; institute pump taxes and import projects to optimize use of ground and surface water; and establish loan programs, conservation rebates, and, in some cases, tiered pricing structures. In short, appurtenant groups of users have, here as elsewhere, voluntarily recognized the need to conserve, and they have advanced creative methods to meet that need, but in a way that emphasizes local control.

Thompson also describes at length the failure of state governments to facilitate voluntary transfers: "Legal restrictions on formal water transfers" pose insurmountable transaction costs for many small water users...." In response, users "have employed local institutions to obtain or exchange water." Institutions provide members with "largely interchangeable water entitlements," allowing the development of standardized markets; members of institutions are also likely to be familiar with each other's needs; and many institutions have constructed extensive physical infrastructure that facilitates the transfer of water from one user to another. Finally, institutions have proved adept at allocating water during times of scarcity: They can "balance their service areas, water supplies, and storage capabilities.... [They] can also more readily calculate and enforce pro rata allocations." In the language of appurtenancy, physical, social, and functional proximity among users facilitates efficient trade and reallocation within the group. Of course, mutuals and, to an even greater extent, districts can exhibit some of the less savory features of appurtenant communities as well. While they vary in size, and while some encompass a large geographic area and a range of users, their success seems due in part to the homogeneous—and sometimes insular—interests of their membership. Many mutuals are "enlivened with a dollop of community spirit." While such community spirit may make informal transfers between members less cumbersome than statutory transfers and encourage the adoption of stronger conservation measures than those imposed by the state, internal community norms also reflect the exclusionary tendencies and insider-outsider mentality that an appurtenant group of users can exhibit. In particular, Thompson notes the resistance of some local institutions to interjurisdictional transfers, such as "ag-urban" trades. Such resistance can often be the result of "parochical self-interest that is inconsistent with broader societal good."

Importantly, however, Thompson does not advocate complete centralization of control as the solution to this problem, and he is wary of stripping too much authority from local institutions and losing the benefits of appurtenancy:

> Despite the potential advantages of creating larger water institutions... attempts to enlarge or consolidate current institutions will inevitably encounter numerous problems and limitations....... Legislatures therefore may do best by merely encouraging institutions to examine the advantages of forming larger units, and then making it as easy as possible for them to do so.

This approach might best be described as more nesting—increased usage of intermediate, umbrella institutions that can facilitate interregional trades among various sets of appurtenant groups, and increased legislative support for such institutions. While Thompson recognizes that more heavy-handed state oversight may in some instances be necessary, his preferred solution is neither purely regulatory nor purely market-based, but is rather institutional. In the multilayered regime he proposes, larger associations can overcome obstacles to market trade that currently exist by producing internal economies of scale and internalizing the costs imposed by one group on others.

Some Policy Suggestions : Having reconceptualized appurtenancy as a principle of functional rather than physical proximity, and having determined that appurtenancy can function within a larger-scale nested regime, we can now begin to ask how such a regime might work in practice in the context of eastern water. The four models outlined above—reverse command-and-control, market-based tradable allowances allocated to communities, horizontal governance, and increased nesting through umbrella organizations—may provide policymakers with a starting point. Berkes, using different terminology, similarly suggests a range of ways in which states can facilitate cross-scale management, including "legitimization of local institutions," enabling legislation that gives authority to "locally devised rules," capacity building "at all levels," and "institution building." What "legitimization" or "institution building" will mean in practice will depend on the particular institutional structure in place in each jurisdiction, and the types of user associations that exist or that might be formed. While a comprehensive policy proposal is beyond the scope of this Note, I would like to offer a few preliminary thoughts as to how policymakers might incorporate appurtenancy into eastern water management.

Given the efficiency and equity concerns posed by purely local organizations, state regulatory agencies should retain ultimate authority over the allocation of water rights. The question, then, is how states might recognize appurtenant groups and include them—formally or informally—in water management decisions. Perhaps the simplest option would be to encourage the development of voluntary user associations, without providing such groups any formal legal recognition. Such an arrangement would allow appurtenant groups to form through self-identification, perhaps utilizing industry or interest groups already in existence. States could offer encouragement in several ways. As Ostrom's study of groundwater basins in southern California demonstrates, provision of a public forum for users can facilitate negotiation, information-sharing, clarification of rights, and adoption of voluntary constraints by users. The state might further support voluntary user action by offering assistance with information-gathering (recall the recordkeeping function provided by the state of Kumaon to local authorities) and the enforcement of any voluntary agreements among users (for example, enforcement of trap limits in Maine lobstering waters, and the provision of one-third of the costs of a monitor for groundwater basin withdrawals in southern California). This approach could be characterized as capacity-building or institution-building, within Berkes's taxonomy, and would in some respects parallel the reverse command-and-control model delineated above. The primary objective would be to aid in the development and enforcement of conservation norms developed voluntarily by those most familiar with the resource.

Beyond such informal facilitation, state regulators could explore a variety of ways to offer appurtenant groups more formal legal recognition. One possibility that could work within a modified version of existing permitting systems would be to issue blocks of permits to groups of users rather than to individuals, along the lines of the market-based tradable allowances model. Appurtenant groups might be defined according to categories of uses (for example, farming, residential water supply, commercial development, recreation). Such an arrangement would afford appurtenant groups a measure of management authority regarding their own usage. Freedom to trade and reallocate rights within groups could assist in the development of conservation measures and efficiency improvements. Appurtenant groups might also prove more effective at distributing water among members during times of scarcity and responding to fluctuations in user need and resource availability.

The existence of a larger market of permits might also facilitate intergroup coordination. At the same time, a centrally administered permitting scheme would preserve the ability of state administrators to set certain policy goals (such as equitable distribution), to monitor overall usage, and to ensure coordination of different user groups. My hope is that the examples of nested regimes and suggestions for eastern water offered in this Section will provide a starting point for the reassessment of regulated riparianism. Regulators should resist the tendency to call for greater centralization and regulatory control, and should question the assumption that ex ante decisions made by an administrative agency can alone solve the problems of scarcity and allocation that trouble eastern water today. The institutional insights gleaned from appurtenancy doctrine should inform the shape of water management institutions going forward.

To a large extent, this Note remains agnostic as to the appropriate degree of state oversight, noting only that administrators should be attentive to the institutional model adopted, as certain regulatory forms may themselves produce inefficiencies. Specifically, regulators might find that cross-group coordination would best be achieved not via more assertive forms of regulation, which have proved disastrous in numerous arenas, but rather by market incentives or the formation of larger umbrella institutions encompassing multiple smaller user groups. On the other hand, a more aggressive regulatory approach may be necessary where parochial community interests produce cross-group inefficiencies or inequities that the market alone cannot resolve. Precisely what balance should be struck between appurtenant groups and state oversight in the management of eastern water is a question the Note does not seek to answer, and one whose answer will very likely be contextual. Having argued for multilevel governance and having sketched a few possibilities, this Note recognizes that the remaining challenge is to understand exactly what institutional design—what allocation of decisionmaking, monitoring, and enforcement functions—will best achieve the goals of efficiency, equity, and conservation in water usage and distribution.

At a time when eastern states face continuing water shortages, and regulated riparian statutes are in need of reassessment, this Note offers new ideas about water management by turning to two different sources of inspiration: the common law and the growing body of literature on common property regimes. Despite the criticism it has garnered, the appurtenancy requirement exhibits certain durable principles for conserving scarce resources that this Note seeks to highlight. Moreover, in comparing appurtenancy to reasonable use, this Note provides a new theoretical perspective contrasting limiting and balancing rules—with which to assess resource management systems and the rules they impose. The Note puts these conclusions to work in its analysis of regulated riparianism, concluding that the retention of a reasonableness inquiry may undermine efforts to conserve scarce water supplies. Rejecting the efficacy of such open-ended balancing tests, the Note instead urges consideration of a new institutional model—the nested enterprise model—as a way to manage scarce resources. In the context of eastern water, the nested enterprise model may offer policymakers a way to incorporate a reconceptualized principle of appurtenancy—a principle that attends to scarcity and respects the expertise of users in conserving common resources—into the broader oversight and authority of the state. While the doctrinal rule may have faded, the core concepts of appurtenancy can thus have continuing relevance. Embedding those concepts in our water law and policy will better allow us to conserve and manage our water supply.

Stockwatering Exemption

Imagine a dairy farm in the arid eastern region of Washington State. The dairy's herd of 2500 cows consumes approximately 45,000 gallons of water per day (gpd). Water is a public resource in Washington, so prospective water users must apply for a permit from the State before withdrawing water from streams or aquifers. The permitting statute exempts specific types of small withdrawals from these permitting requirements, however, including "any withdrawal of public ground waters for stock-watering purposes...." Although other exempt uses are limited to 5000 gpd, the dairy farm believes that this limitation does not apply to its withdrawal of 45,000 gpd because the statute exempts "any withdrawal" for stock-watering purposes.

The Washington State Department of Ecology (Ecology), the Pollution Control Hearings Board (PCHB), and the Washington State Attorney General (Attorney General) recently examined this issue and arrived at conflicting conclusions. Ecology initially concluded that stock-watering withdrawals are limited to 5000 gpd by relying on the statutory and historical context of the exemption. In 2001 the PCHB agreed with this interpretation in DeVries v. Department of Ecology. A 2005 Attorney General opinion concluded that the plain meaning of the statute allows unlimited stock-watering withdrawals, however, and Ecology has subsequently changed its practices to accord with this opinion.

The distinction between a limited and unlimited exemption is significant because declining groundwater levels and a proliferation of exempt groundwater withdrawals currently impede sustainable management of Washington's groundwater resources. Increasing reliance on groundwater has overstressed aquifers across the nation. In Washington, rapid population growth has placed significant pressure on groundwater resources.

These demands have caused declining groundwater levels across the state, especially in the eastern and southeastern regions, where the majority of the state's livestock farms are located. As water availability declines, Ecology is forced to deny an increasing number of water use applications to protect senior water rights. Consequently, new water users are turning to the statutory exemptions from the groundwater permitting requirements. Although the exact quantity of exempt withdrawals is unknown, because such withdrawals are unregulated, it is clear that the total amount of groundwater withdrawn under the exemption is significant. An unlimited exemption for stock-watering could exacerbate current groundwater management issues, because such exempt withdrawals occur outside Ecology's permitting authority. This Comment analyzes the conflicting interpretations of Washington's stock-watering exemption and argues that the exemption is limited to 5000 gpd. This conclusion is based on Washington's principles of statutory interpretation, the statute's historical background, and public policy considerations.

Washington Water Law Comprehensively Regulates Groundwater

Washington groundwater is comprehensively regulated by a groundwater code that exempts small withdrawals from its permitting system. The Legislature enacted the groundwater code to prevent destructive competition for groundwater by providing uniform regulations and a centralized administration for governing water rights. Although specific types of withdrawals are exempt from the groundwater code's permitting requirements, they remain subject to other regulations within the groundwater code.

The Surface and Groundwater Codes Were Enacted to Protect Senior Water Users, Minimize Disputes Over Water Rights, and Protect Public Water Resources

The prior appropriation doctrine governs water use in Washington, as it does in most western states. Under the riparian doctrine, which traditionally governed water use in the eastern states, only lands adjacent to water are entitled to a water right. In contrast, the prior appropriation doctrine allows water users to acquire water rights by appropriating water and putting it to a beneficial use. Prior appropriation also protects senior water users by providing that between appropriators from the same water source, the first in time has a higher priority water right. The gold miners and farmers who first settled the western territories developed this system because land that was not adjacent to water in the arid regions of the west was practically worthless under the riparian doctrine. Appropriation based on need, rather than on riparian property rights, facilitated successful settlement.

Between 1889 and 1917, Washington recognized water rights under both the riparian and prior appropriation doctrines. Beginning with Thorpe v. Tenem Ditch Co. in 1889, the Washington State Supreme Court considered the question of water rights in several cases and protected both rights by appropriation and rights by riparian ownership. In addition to the case law, Washington had many conflicting statutes regarding the right to use water. This resulted in a chaotic and confused state of the law, which, in turn, led to continuous disputes among water users throughout the state. To create clarity and security for water users, the Washington State Legislature enacted a comprehensive surface water code in 1917. This code adopted the prior appropriation doctrine and created a centralized administration for determining the existence of water rights and establishing new rights, thereby ensuring the most efficient use of water resources.

Several decades later, the 1945 Legislature passed the groundwater code as a supplement to the surface water code, extending the protections and regulations of the surface water code to groundwater. Although the Legislature did not explicitly state its purpose for enacting the groundwater code, the Washington Department of Conservation and Development (Department of Conservation), which was the agency responsible for water resources in 1945, described the reasons for its enactment in a 1946 report. In the report, the Department of Conservation stated that the groundwater bill was prepared and sponsored at the request of an association of Washington cities to protect the 126 cities that depended on groundwater for domestic uses, irrigation, and industrial purposes.

The Department of Conservation also noted that "destructive competition has been the experience of certain other states," and "in some sections of this State, uncontrolled withdrawal of [ground]water has already caused damage to existing rights and investments." This report indicates that the Legislature enacted the groundwater code to protect existing rights, avoid damaging competition, and provide security for future development.

The Ground-water Code Provides Comprehensive Regulations for Obtaining a Water Right Permit, with an Exemption for Small Withdrawals of Water

The groundwater code is supplementary to the surface water code. Because the groundwater code extended the provisions of the surface water code to govern groundwater appropriations, it is important to first understand the relevant provisions of the surface water code. The surface water code adopted the prior appropriation doctrine and created a permitting system that allows the state to control the distribution and utilization of surface water resources. Under this permitting system, a prospective water user must obtain a permit from Ecology before appropriating water. Before issuing a permit, Ecology must find "(1) that water is available, (2) for a beneficial use, and that (3) an appropriation will not impair existing rights or (4) be detrimental to the public welfare." The surface water code also mandates that none of its provisions "shall be construed to lessen, enlarge, or modify... any existing right acquired by appropriation, or otherwise." The groundwater code adopted most of these surface water code provisions. The groundwater code's purpose is "regulating and controlling the ground waters of the state." Because groundwater withdrawals can impact surface water resources in areas where surface water and groundwater are connected, the groundwater code protects existing surface water rights by mandating that surface water rights existing when the code was enacted "shall not be affected or impaired by any of the provisions of this supplementary chapter." The groundwater code also adopts the prior appropriation rules and permitting system established in the surface water code. Unlike the surface water code, however, the groundwater code provides an exception to the permitting system.

The groundwater code's general permitting requirement provides: After June 6, 1945, no withdrawal of public ground waters of the state shall be begun, nor shall any well or other works for such withdrawal be constructed, unless an application to appropriate such waters has been made to the department [of Ecology] and a permit has been granted by it as herein provided....

This general rule requires prospective groundwater users to apply to Ecology for a permit before withdrawing public groundwater, and Ecology must make the same findings required in the surface water code before granting the permit application. This general rule is followed by an exception, in the form of an exemption provision, and two modifying provisions. The exemption provision states:

Except, however, That any withdrawal of public ground waters for stock-watering purposes, or for the watering of a lawn or of a noncommercial garden not exceeding one-half acre in area, or for single or group domestic uses in an amount not exceeding five thousand gallons a day, or as provided in RCW 90.44.052 [applicable only in Whitman County], or for an industrial purpose in an amount not exceeding five thousand gallons a day, is and shall be exempt from the provisions of this section, but, to the extent that it is regularly used beneficially, shall be entitled to a right equal to that established by a permit issued under the provisions of this chapter....

This provision exempts four categories of water use from the general permitting rule: stock-watering, irrigation, domestic uses, and industrial uses. Three of these categories are specifically limited-irrigation of a lawn or garden is restricted to a half acre, and domestic and industrial uses are limited to 5000 gpd-while the stock-watering provision does not contain a specific limitation. Although these categories of withdrawals are exempt from the permitting requirement, all groundwater withdrawals are subject to the mandate that existing surface water appropriations "shall not be affected or impaired by any provisions of [the groundwater code]," and the mandate that senior appropriators are entitled to a "safe sustaining yield" of groundwater.

Two modifying provisions follow the exemption provision: Provided, however, that the department from time to time may require the person or agency making any such small withdrawal to furnish information as to the means for and the quantity of that withdrawal: provided, further, that at the option of the party making withdrawals of ground waters of the state not exceeding five thousand gallons per day, applications under this section or declarations under RCW 90.44.090 may be filed and permits and certificates obtained in the same manner and under the same requirements as is in this chapter provided in the case of withdrawals in excess of five thousand gallons a day. The first proviso authorizes Ecology to require exempt users to report the purpose and quantity of their withdrawals, and describes all exempt uses as "small withdrawal[s]." The second proviso, added in 1947, provides that users may voluntarily obtain a permit for their exempt withdrawals, and it describes the exempt withdrawals as "not exceeding five thousand gallons per day." When reviewing the statute as a whole, it is unclear whether the statute allows unlimited use because it lacks a specific limitation for the stock-watering exemption, or whether the statute limits stock-watering to "small withdrawals" not exceeding 5000 gpd through the modifying provisos.

While the groundwater code's stock-watering exemption lacks clarity, what is clear is that the circumstances surrounding the surface water and groundwater codes' enactment strongly indicate that the Legislature created these codes to minimize water disputes and protect the public's water resources from destructive competition. The groundwater code adopted the surface water code's protections and regulations for groundwater. Unlike the surface water code, however, the groundwater code contains an exception to the general permitting requirements for specific categories of water use, including withdrawals for stockwatering. This exemption statute fails to clearly convey the limitations that govern the stock-watering exemption.

The Legislature Enacted the Exemption in 1945

Although the exemption statute does not specifically limit stockwatering, historical context shows that stock-watering did not exceed 1500 gpd on an average farm in Washington when the

Legislature enacted the statute in 1945. Since 1945, however, the industrialization of the livestock industry has substantially increased the size of livestock farms and the amount of water a single farm is capable of using. The historical background and subsequent development of stock-watering provides a historical context for understanding the Legislature's intent when it enacted the stock-watering exemption.

Stock-watering Requirements For a Fully Developed Family Farm in 1945 Did Not Exceed 1500 Gallons Per Day

Historical circumstances surrounding the stock-watering exemption's enactment reveal that an average family farm required 1500 gpd or less in 1945. At the time the Legislature enacted the statute, Washington and the United States Bureau of Reclamation were attempting to populate the Columbia Basin region with family farms. The Columbia Basin Project was located below the newly constructed Grand Coulee Dam, in the dry eastern region of the state. It was the Bureau of Reclamation's largest project, part of its plan to "develop the West through the creation of permanent family farms on Federal Reclamation projects." The Bureau of Reclamation expected the Columbia Basin Project to strengthen the agricultural economy of the Pacific Northwest once post-World War II settlers developed the project area with irrigation water provided by the Grand Coulee Dam. For settlement to succeed, every rural settler needed a domestic supply of water at a minimum cost.

A 1945 Bureau of Reclamation report on farm improvement recommended that the supply of domestic water "should be sufficient (1) to satisfy the personal demands of the settlers, including the operation of plumbing facilities; (2) to water livestock; (3) to sprinkle lawns and small gardens occasionally; (4) to process farm products; and (5) to provide some fire protection." These recommended categories parallel the categories codified in Washington's groundwater exemption statute in the same year that the Bureau of Reclamation published its report. The report also noted that because climate and topography in the Columbia Basin limited the use of ditches, canals, rivers, and creeks as water sources, groundwater was the most promising source for rural development. The farm improvement report also advised that "total daily requirements of the average farm may be only 200 gallons during the early years" and "will expand to perhaps 1,500 gallons during the mature development." This estimation is consistent with the 1940 Washington Census of Agriculture statistics, which reveal that Washington farms at this time were typically small farms of 100 acres or less. One year after the Legislature enacted the groundwater code, the Department of Conservation published an overview of the groundwater code, which described the groundwater exemption statute as:

Individual Domestic Supply Exempt

The Ground Water Code exempts from administrative control the withdrawal of public ground water for any purpose where the quantity is less than 5,000 gallons per day. This exemption was provided to relieve the small water user of the formalities and costs of obtaining water for his household and domestic needs. Five thousand gallons per day will supply ample water for household use for a family, their garden and lawn irrigation, and stock water.

The Department of Conservation interpreted the statute as limiting all exempt uses, including stock-watering, to 5000 gpd. The agency believed the statute's purpose was to supply "small water users" with water, and the agency considered 5000 gpd an "ample" amount for such users. The Department of Conservation's interpretation limiting all exempt uses to 5000 gpd was consistent with the Bureau of Reclamation's estimation that maximum water use for small family farms would not exceed 1500 gpd.

The Industrialization of American Agriculture Since 1945 Has Significantly Altered the Structure and Concentration of the Livestock Industry

Since 1945, the livestock industry has experienced rapid industrialization and concentration. In recent decades, technological advances and horizontal integration have dramatically altered the way farms are operated, replacing the small-scale family farms that historically typified American agriculture with industrialized farming. Technological improvements have steadily increased farming production efficiency since the 1930s, but when the Legislature enacted the groundwater code in 1945, many still did not believe that large-scale farming would ever replace the family farm as a staple of American agriculture. For instance, the Bureau of Reclamation's mission was to "create and sustain family farms in the West" by providing water with reclamation projects, like the Grand Coulee Dam and the Columbia Basin Project. The vision of a rural West populated by small farms persisted from the 1930s into the 1960s, even though this vision competed with emerging agricultural businesses and growing cities that "created new opportunities for those eager to flee the countryside." Congress did not raise the acreage limitation on reclamation farms until 1982, when it raised the limitation from 160 acres to 960 acres. This action convinced observers that the Bureau of Reclamation had finally abandoned its original family farm mission and endorsed agribusiness.

As technology increasingly advanced, engineered livestock confinement facilities became widespread between the 196Os and 198Os, replacing traditional pasture methods and leading to a concentration of livestock on industrialized farms. From the 1950s to the 1990s, the number of farms in the United States decreased by fifty percent, and concentrated livestock operations more than doubled between 1982 and 1997. As recently as 1987, a typical dairy herd had fewer than seventytwo cows, but now large dairy farms with 500 or more cows are the fastest growing segment of the dairy industry in the western states.

Washington agriculture statistics also reflect this trend of declining farms numbers and increasing farm size. In 1940, Washington State had approximately 81,600 farms, and the majority of those farms were 100 acres or less in size. The total number of livestock was approximately 1,500,000, with cattle representing about half that number and sheep comprising most of the other half. In contrast, by 2002 Washington had approximately 36,000 farms with an average size of 430 acres. Census statistics classified about one third of the farms as livestock industries. Additionally, the total number of livestock in 2002 amounted to approximately 1,300,000, including 1,100,000 cattle. As these statistics reveal, livestock numbers have not significantly decreased since the 1940s, but they have become concentrated on half the number of farms. At the same time, livestock demographics have shifted to predominantly cattle.

This shift in demographics is significant because of the difference in the amount of water consumed by cows and smaller livestock. In a temperate climate, beef cattle consume about twelve gallons of water per day, and dairy cows consume about thirty-five gallons. In contrast, sheep and goats consume one to four gallons per day. An increasing concentration of cows leads to higher daily water consumption. For example, in 2006 Washington State had approximately 490 dairy farms, with an average herd size of 475 cows per farm. If each dairy cow consumed thirty-five gallons of water per day, the total average daily consumption per dairy farm was approximately 16,625 gpd. In short, when Washington's Legislature enacted the groundwater code in 1945, the state was encouraging family farm development in the Columbia Basin and needed to provide water for settlers at minimum cost. At that time, the average fully developed farm required less than 1500 gallons of water per day. One year after the exemption statute was enacted, the Department of Conservation concluded that it limited all withdrawals to an "ample" amount of 5000 gpd. Since the statute's enactment, however, the industrialization of the livestock industry has decreased the number of farms in Washington and increased farm size, concentrating a greater number of livestock onto fewer farms and thereby increasing the amount of water a single livestock farm is capable of using.

Conflicting Interpretations of the Stockwatering Exemption

Ecology, the PCHB, and the Attorney General have interpreted the stock-watering exemption and arrived at conflicting conclusions. The agencies dispute whether the exemption is limited or unlimited. Ecology originally limited the stock-watering exemption to 5000 gpd based on the overall structure of the groundwater code and the statute's historical context. The PCHB upheld this interpretation in a dispute between a dairy farm and Ecology in 2001, but in 2005, the Attorney General issued an opinion concluding that the stock-watering exemption is unlimited. Ecology subsequently changed its practices to conform to the Attorney General's opinion.

The Department of Ecology Originally Interpreted the Stockwatering Exemption as Limited to 5000 Gallons Per Day

Ecology, the agency responsible for administering the groundwater code, originally interpreted the stock-watering exemption as limited to 5000 gpd. The agency based its interpretation on the language of the two provisos that describe all exempt uses as "small withdrawals" "not exceeding five thousand gallons per day." Also, the groundwater code specifically states that no existing surface water rights "shall be affected or impaired by any of the provisions of this supplementary chapter." Ecology reasoned that this indicates the groundwater exemption provision is not unlimited, because unlimited withdrawals could potentially impair senior surface water users. Ecology also relied on the Department of Conservation's 1946 interpretation that stock-water withdrawals were limited to 5000 gpd, and a 1942 report on domestic water supply by the Washington State Planning Council indicating that total farm demand for domestic and stock-water was expected to be 1500 gpd. For these reasons, Ecology concluded that stock-watering withdrawals are limited to 5000 gpd under the exemption statute and large livestock operations, such as cattle feedlots and commercial dairies, may not use the statute to withdraw unlimited groundwater for stock-watering purposes.

The PCHB Also Interpreted the Stock-Watering Exemption as Limited to 5000 Gallons Per Day

In a 2001 case, the PCHB agreed with Ecology that the stockwatering exemption is limited to 5000 gpd. In DeVries v. Department of Ecology, a dairy farm relied on the stock-watering exemption to withdraw between 39,000 to 56,000 gpd for its herd of 2261 dairy cows. The herd was expected to eventually reach 4400 cows, consuming about 110,000 gallons of water per day. When Ecology ordered the dairy to restrict its groundwater use to 5000 gpd until it obtained a permit to withdraw more, DeVries appealed to the PCHB, arguing that the plain language of the exemption allows unlimited withdrawals for stock-watering. The PCHB concluded that the stockwatering exemption is limited to 5000 gpd.

The PCHB reasoned that the first proviso indicates the Legislature intended to limit all exemptions to "small withdrawals," and the second proviso defines small withdrawals as "not exceeding five thousand gallons per day." The PCHB also reasoned that canons of statutory construction support this interpretation. The PCHB described the general permitting requirement as "the central purpose of the 1945 groundwater code." Under the rule of narrowly construing exceptions to give full effect to a general provision, the PCHB concluded that it should narrowly construe the stock-watering exemption to give full effect to the general permitting requirement. Furthermore, under the rule of avoiding literal readings if they produce absurd consequences, the PCHB concluded that "[a]ll of the objectives of the groundwater code would be undermined if the stockwatering exemption is for an unlimited quantity," because "an unlimited, and uncontrollable, potential for withdrawal of groundwater" could potentially have a "devastating effect on the ability of the state to protect the senior water right holders or to grant future water rights."

The Attorney General Interpreted the Stock-watering Exemption as Unlimited and Ecology Subsequently Changed its Interpretation

The Attorney General issued a formal opinion in 2005 interpreting the stock-watering exemption as unlimited. When Washington cattle ranchers reported being turned down for financing because of the 5000 gpd limit Ecology was imposing on exempt stock-watering withdrawals, two legislators from eastern Washington requested an opinion from the Washington State Attorney General. Although Attorney General opinions are advisory in character and do not bind state officers, "officials are entitled to [the Attorney General's] opinion before they do anything as they are not protected [from civil liability] if acting without the advice of their duly constituted legal adviser." If an official disregards an Attorney General opinion, "they do so at their own peril."

Because the stock-watering provision is the only exemption that is not immediately qualified by a limitation, the Attorney General concluded that the plain language of the statute does not limit stock-watering withdrawals based on ordinary rules of language and grammar. While the Attorney General acknowledged Ecology's argument that an unlimited exemption is "inherently inconsistent with the general policy of requiring permits for groundwater withdrawals in order to provide for an orderly and consistent administration of an important and limited public resource," he noted that Ecology could alternatively require information on the use of exempted withdrawals for use in its administrative and enforcement decisions.

Shortly after the Attorney General issued this opinion, Ecology issued a press release stating that even though it "had consistently interpreted the livestock watering exemption as limited to 5,000 gallons a day," the agency would change its practices to conform to the new legal guidance provided by the Attorney General, and "will not take any action to limit groundwater use under the stock-water exemption to 5,000 gpd."

The Primary Goal of Statutory Interpretation

The stock-watering exemption must be interpreted in accordance with Washington's principles of statutory interpretation. A court's fundamental objective in statutory interpretation is to implement the Legislature's intent. Courts first analyse the plain meaning of a disputed statutory provision based on its text and the context of related statutes. A "plain" meaning based on literal or express wording may be restricted, however, if it does not implement the Legislature's intent and impairs the spirit or purpose of an enactment. If the statute remains ambiguous after conducting a plain meaning analysis, then courts may resolve the ambiguity by examining a statute's legislative history and the circumstances surrounding its enactment to determine legislative intent. When a statute is ambiguous, courts may also consider the public policies underlying the statute.

Washington Courts First Analyse the Plain Meaning of a Statute to Ascertain the Legislature's Intent

A court's fundamental objective in statutory interpretation is ascertaining and implementing the Legislature's intent. The first step in statutory interpretation is to look to the text of the statute, and if the statute's meaning is plain, then courts must give effect to that plain meaning as an expression of the Legislature's intent. Although Washington courts previously looked only to the disputed statutory provision when analysing the plain meaning of a statute, the Washington State Supreme Court recently expanded the analysis in Department of Ecology v. Campbell & Gwinn to include "all that the Legislature has said in the statute and related statutes which disclose legislative intent about the provision in question." Washington's expanded plain meaning rule mandates that courts consider every word within the text of the disputed statute, as well as the context of related statutes and the integrity of the overall statutory scheme. For example, where a disputed provision is an exception to a general rule, context suggests that a court should interpret the exception narrowly.

Although courts often do not clearly explain how they have arrived at their conclusion that a statute's meaning is plain, their rationale becomes clearer when the reader remembers that a court's primary goal in statutory interpretation is ascertaining and implementing the Legislature's intent. For this reason, courts avoid literal interpretations if they do not implement the Legislature's intent. A statute's plain meaning should not be based on "literal" wording if it produces "[s]trained, unlikely, or absurd consequences." Rather, "the purpose of an enactment should prevail over express but inept wording." Once a court has ascertained the Legislature's intent, "general words [within the statute] may be restrained to [the general intent], and those of narrower import may be expanded to effectuate that intent."

For example, in Department of Ecology v. Campbell & Gwinn the Washington State Supreme Court applied the plain meaning rule to the groundwater exemption statute addressed by this Comment. The statutory provision at issue was the domestic use exemption, which exempts "any withdrawal... for single or group domestic uses in an amount not exceeding five thousand gallons a day." The court was asked to determine whether a residential subdivision developer could use the exemption to drill multiple wells that would individually withdraw less than 5000 gpd but would collectively withdraw more than 5000 gpd.

The court considered whether the plain meaning of the words "any withdrawal" in the exemption statute is "every or all removals of water," and determined that "[n]either term, in and of itself, defines the scope of the right... and the words 'any withdrawal' do not establish the plain meaning of the exemption." When the parties disputed the potential impact of interpreting the exemption to apply to each individual well, the court stated that "[t]he question is more basic, i.e., whether the Legislature even contemplated the possibility that developments of the size in this case, or even larger, would be entitled to exempt withdrawals of 5,000 gpd for each of their lots." The court concluded that "[g]iven the limitation on [domestic] uses, and the overall goal of regulation to assure protection of existing rights and the public interest, it is clear that the Legislature did not intend that possibility when this statute was enacted." The court went on to hold that the plain meaning of the domestic use exemption, as determined by the context of related statutory provisions, limited the developer to one 5000 gpd withdrawal for the entire project. As Campbell & Gwinn demonstrates, the plain meaning analysis requires a court to ascertain and implement the meaning that the Legislature intended when the statute was enacted.

Washington Courts May Consider Legislative History, Historical Context, Agency Interpretations, and Policy Considerations if a Statute is Ambiguous

If a statute remains susceptible to more than one reasonable meaning under the plain meaning analysis, then courts may look to sources outside of the statute's text to ascertain the Legislature's intent. Washington courts have held that ambiguity exists when multiple interpretations are reasonable, not merely conceivable. In other words, ambiguity exists when a reasonably well-informed person is capable of understanding a statute in two or more different senses.

Where a statute is ambiguous, Washington courts may look to the legislative history of a statute, the circumstances surrounding its enactment, and contemporaneous interpretations of the statute to determine legislative intent. Legislative history includes hearings, committee reports, and floor debates. Historical context includes circumstances leading up to and surrounding a statute's enactment.

Courts look to the historical context in which the Legislature enacted a statute to identify the problem the Legislature intended the statute to solve and ascertain the Legislature's purpose for enacting the statute. Even where a statute is not ambiguous, the Washington State Supreme Court has used historical circumstances to support the plain meaning of the statute. For instance, in Restaurant Development Inc. v. Cananwill Inc., the court determined that the plain meaning of the Insurance Premium Finance Company Act instructs insurance companies to calculate service charges by using an "add-on" method, and then discussed circumstances surrounding the enactment of the statute in 1969 to support its plain meaning interpretation.

When a statute is ambiguous, courts may also defer to contemporaneous agency interpretations of the statute. The reason for this is not because the agency interpretation is evidence of legislative intent, but because Washington courts have a policy of deferring to agency interpretations when (1) a statute is ambiguous, (2) the agency is charged with enforcing the statute, and (3) the agency has special expertise in the statute's subject matter. Attorney General interpretations are also accorded "considerable weight" if a statute is ambiguous, although the Washington State Supreme Court recently stated that it "gives[s] little deference to Attorney General opinions on issues of statutory construction." Washington courts have not indicated which interpretation receives greater deference when the Attorney General and an agency disagree.

Finally, where a public interest is affected, an interpretation that favours the public and implements the public policy underlying the statutory provision is preferred. Public policy is found by examining the history, purpose, language, and effect of the statute. Generally courts only use public policy to resolve ambiguous statutes because of the indefinite nature of "public policy." Washington courts have stated that a "paramount concern" when interpreting a statute is ensuring that "the interpretation is consistent with the underlying policy of the statute." Courts cannot amend statutes that changing societal conditions have rendered obsolete. Rather, where two reasonable interpretations of a statute exist, a court should adopt the interpretation that better advances the overall legislative purpose and policy behind the statute.

For example, in In re Hastings the Washington State Supreme Court said, "In passing on the meaning of any legislative enactment, we look to the language of the statute [and] the policy behind it." The court then examined the history of Washington's pretermitted heir statute and interpretations of similar statutes in other jurisdictions to determine the policy underlying the act. Once the Hastings court determined that the underlying policy was to protect heirs who are omitted from a will due to the failing mind of an elderly testator, the court held that it must "construe [the statute] in a way that is 'consonant with the obvious purpose'" and declined to extend the statute to include children who were omitted because they had predeceased the testator.

Overall, a court's fundamental objective in statutory interpretation is to ascertain and implement the Legislature's intent. Washington courts first attempt to ascertain this intent by analysing the plain meaning of a statute. If the statute is ambiguous, Washington courts may examine the legislative history of a statute, historical circumstances surrounding its enactment, and contemporaneous interpretations to determine the Legislature's intent. Courts may also defer to agency interpretations or examine a statute's underlying public policy when interpreting an ambiguous statute.

Washington's Stock-watering Exemption Is Limited To 5000 Gallons Per Day

The proper interpretation of the stock-watering exemption begins with a plain meaning analysis of the statute, bearing in mind that the ultimate goal is ascertaining and implementing the Legislature's intent. The text of the exemption statute and the context of related statutory provisions within the groundwater code reveal that the enacting Legislature did not intend to create an unlimited stock-watering exemption. Rather, the Legislature instead intended to limit exempt withdrawals of groundwater to 5000 gpd. Even if the text and context of the statute did not clearly support limiting the exemption, the historical circumstances surrounding the statute's enactment and the groundwater code's underlying public policy further support limiting stock-watering withdrawals to 5000 gpd.

The Text and Context of the Exemption Statute Indicate That the Legislature Intended to Limit the Stock-watering Exemption to 5000 Gallons Per Day

The stock-watering exemption is a limited exemption according to the plain meaning of the statute. Although the text of the stock-watering provision indicates the exemption is unlimited when standing alone, the context of related statutes support narrowly construing the exemption. Furthermore, the context of the exemption statute shows that 5000 gpd is the proper quantity for the limitation.

The Text of the Stock-Watering Provision Supports an Unlimited Interpretation when Considered in Isolation : A plain meaning analysis of the statute begins with the text of the stock-watering exemption, which states that "any withdrawal of public ground waters for stock-watering purposes... shall be exempt from the provisions of this section." If the analysis were limited to this provision, it would be easy to conclude that the statute exempts unlimited withdrawals for livestock based on the ordinary meaning of the words "any withdrawal." However, the Washington State Supreme Court has already considered whether the phrase "any withdrawal" within the exemption statute defines the scope of an exemption, and in Campbell & Gwinn the court held that the words did not define the scope of the domestic use exemption. The expanded plain meaning rule that was adopted in Campbell & Gwinn allowed the court to look beyond the disputed statutory provision and rely on the context of the entire statute and related statutory provisions to determine the proper scope of the domestic use exemption. Since the Campbell & Gwinn court was interpreting the same phrase-"any withdrawal"-within the same statute, courts should apply the same interpretation to the stock watering exemption. Thus, courts must discern the scope of the stock-watering exemption from the text and context of the entire statute, and cannot define the statute's scope solely by the ordinary meaning of the words "any withdrawal."

The Context of the Exemption Provision Also Supports an Unlimited Interpretation of the Stock-Watering Exemption : The first level of context for the stock-watering provision is the context of the entire exemption provision. The exemption provision provides that any withdrawal for (1) stock-watering, (2) lawn and noncommercial garden irrigation one half acre or less, (3) single or group domestic uses 5000 gpd or less, or (4) industrial uses 5000 gpd or less, is exempt from the general permitting requirement. Because stock-watering is the only exempt withdrawal that is not immediately qualified by a limitation, it would appear that the Legislature intended stock-watering withdrawals to be unlimited. However, courts should avoid literal readings if they lead to absurd consequences. In a case such as DeVries v. Department of Ecology, allowing a dairy farm to withdraw up to 100,000 gpd under the stock-watering exemption is an absurd result based on the dramatic disparity between such a withdrawal and the other exempt withdrawals, which are limited to one half acre of irrigation or 5000 gpd. To resolve the inconsistency between the structure of the exemption provision, which appears to allow unlimited withdrawals for stock-watering, and the absurd results produced by such an interpretation, it is necessary to examine the broader context of the entire exemption statute and other statutes within the groundwater code.

The Context of the Entire Exemption Statute and the Groundwater Code Support a Limited Interpretation of the Stock-Watering Exemption : The context of the entire exemption statute and other statutes within the groundwater code indicate that exempt stock-watering withdrawals are not unlimited. The overall structure of the exemption statute reveals that it is primarily a general rule requiring all prospective water users to obtain a permit before withdrawing groundwater.

The exemption provision follows this general rule, and two provisos modify the exemption provision. The exemption provision begins with the words "EXCEPT, HOWEVER...." indicating that it is an exception to the preceding rule.208 Accepted principles of statutory interpretation mandate that courts narrowly interpret exceptions to general rules. Furthermore, the two modifying provisos that follow the exemption provision characterize all of the exempt withdrawals as "small withdrawals" that do not exceed 5000 gpd.

In this way, the structure and text of the entire statute support the conclusion that the Legislature only intended to exempt small withdrawals from the general permitting requirement, and courts must narrowly construe such exemptions to give maximum effect to the general permitting rule. An unlimited stockwatering exemption would be inconsistent with the Legislature's intent that water users must obtain a permit before withdrawing groundwater in most circumstances, especially considering the comparatively small quantities of water allowed under the other exempt uses and the Legislature's description of the exempt uses as "small withdrawals]" "not exceeding five thousand gallons a day.""

Related statutes within the groundwater code further reinforce the interpretation that stock-watering is a limited exemption. To begin with, advocates of an unlimited stock-watering exemption argue that an unlimited exemption is not inconsistent with the overall structure of the groundwater code because there are other enforcement mechanisms within the groundwater code. The exemption statute itself provides that Ecology may require a person making an exempt withdrawal to provide information about the quantity of their withdrawal, and a related statute authorizes Ecology to limit groundwater withdrawals to protect senior appropriators. The Washington State Supreme Court addressed this argument in Campbell & Gwinn, however, and the court concluded "after-the-fact remedies will not serve legislative purposes as effectively as review before [groundwater] appropriation occurs," because "damage will already have been done" by the time the enforcement mechanisms are invoked.

Furthermore, the stated purpose of the groundwater code is "regulating and controlling ground waters of the state." The groundwater code also mandates that senior surface water rights "shall not be affected or impaired by any of the provisions of this supplementary chapter," and provides that senior appropriators "shall enjoy the right to have any withdrawals by a subsequent appropriator of ground water limited to an amount that will maintain and provide a safe sustaining yield" of groundwater for prior appropriators.

The State cannot properly regulate and control groundwater and protect senior appropriators if unlimited withdrawals of groundwater are made outside the regulatory permitting scheme. An exemption allowing unlimited groundwater pumping could have a detrimental effect on surface water flows that are connected to groundwater sources, and a "devastating effect on the ability of the state to protect the senior water right holders or to grant future water rights." In DeVries, the PCHB concluded that "[a]ll of the objectives of the groundwater code would be undermined if the stockwatering exemption is for an unlimited quantity." As these statutes demonstrate, an unlimited exemption is inconsistent with the groundwater code's overall purpose, which is to protect senior appropriators and regulate withdrawals to maintain a safe and sustaining yield of groundwater.

The Proper Limitation for the Stock-Watering Exemption is 5000 Gallons Per Day Based on the Context of the Entire Exemption Statute : The text and context of the statute indicate that the Legislature intended to create a limited stock-watering exemption. The appropriate scope of that limit is 5000 gpd based on the context of the entire exemption statute. Two of the other exempt uses are limited to 5000 gpd, indicating that the Legislature believed this was an appropriate limitation for exempt withdrawals. Furthermore, the first proviso describes the scope of all of the exemptions as "small withdrawal[s]" and the second proviso describes the exemptions as withdrawals that do not exceed 5000 gpd. When read together, these two provisos indicate that the Legislature intended all exempt withdrawals to be small withdrawals that do not exceed 5000 gpd.

The second proviso states that a party withdrawing groundwater "not exceeding five thousand gallons per day" may obtain a permit for their exempt withdrawal in the same manner as a party "withdrawing] in excess of five thousand gallons a day" under the general permitting provisions. It could be argued that the second proviso only applies to the exemptions that specify a 5000 gpd limitation, rather than describing all exempt withdrawals as limited to that amount.

However, a permit provides protection for senior appropriators because it documents the date the water user acquired the water right and establishes the right as senior to all later appropriations. Presumably the Legislature would not allow some exempt users to obtain permits, while denying others the privilege simply because they are using their groundwater withdrawals for a different purpose. There is no logical reason for the Legislature to only allow domestic and industrial withdrawals (exemptions that specify a 5000 gpd limitation) to obtain the protections offered by a permit, but not stock-watering and irrigation withdrawals. Furthermore, this interpretation ignores the first proviso, which describes all of the withdrawals as "small," indicating that they are not unlimited.

In short, a plain meaning analysis of the stock-watering exemption supports interpreting the exemption as limited, rather than unlimited, based on the context of the statute and related statutory provisions. The proper scope of the limitation is 5000 gpd, based on the context of the modifying provisos and other exemption provisions within the statute.

Historical Circumstances Support Limiting the Stock-watering Exemption to 5000 Gallons Per Day

The historical circumstances surrounding the groundwater code's enactment also support limiting the stock-watering exemption to 5000 gpd. Even though the exemption is limited to 5000 gpd according to the plain meaning of the statute, considering historical circumstances is still appropriate in this case. While historical context may not be used to contradict the plain meaning of an unambiguous statute, the Washington State Supreme Court has used historical context to confirm and support the plain meaning of an unambiguous statute. Pre-enactment historical circumstances show that stock-watering was minimal and did not exceed 1500 gpd when the statute was enacted in 1945. Post-enactment developments show that in 1946 the statute was interpreted as limiting all exempt withdrawals to 5000 gpd, and in 1947 the Legislature implicitly endorsed this interpretation when it amended the statute. The correct interpretation of the stock-watering exemption requires an answer to the same fundamental question addressed in Campbell & Gwinn: Did the Legislature even contemplate the possibility that future groundwater users would be entitled to exempt withdrawals that exceed 5000 gpd for stock-watering? Because little legislative history for the groundwater code exists, historical circumstances immediately surrounding the enactment of the groundwater code provide especially valuable insights into the Legislature's intent.

Pre-enactment historical circumstances reveal that the groundwater exemption was almost certainly enacted to support the development of small family farms that were not expected to require more than 1500 gpd. When the Legislature enacted the exemption in 1945, the state and federal government were in the process of developing the Columbia Basin region into family farms. A report from the Bureau of Reclamation issued in the same year that the statute was enacted advised that the average fully developed farm would require less than 1500 gallons of water per day.

The Legislature likely relied on this report when enacting the exemption statute for several reasons. First, the report indicated that securing a domestic supply of water at minimum cost was necessary for successful settlement. The Legislature accomplished this goal by providing an exemption to the general permitting regulations second, the report suggested groundwater was the most promising source for rural development, and the Legislature created an exemption within the groundwater code, but not the surface water code. Third, the necessary categories of water discussed in the report-irrigation of lawns and small gardens, domestic uses, water for livestock-parallel the categories that the Legislature exempted in the groundwater code.

Post-enactment historical circumstances further support the conclusion that the Legislature intended stock-watering to be limited to 5000 gpd. In 1946, one year after the groundwater code was enacted, the Department of Conservation interpreted the statute as allowing withdrawals "for any purpose where the quantity is less than 5,000 gallons per day," and in the agency's opinion, that "[five thousand gallons per day will supply ample water for household use for a family, their garden and lawn irrigation, and stock water." In 1947, one year after this interpretation was published in a report to the Legislature, the Legislature added the second proviso describing all exempt uses as "not exceeding five thousand (5,000) gallons a day." The Legislature was presumably aware of the Department of Conservation's interpretation when it amended the exemption statute, so the Legislature implicitly affirmed the interpretation that the statute limits all exempt uses to 5000 gpd when it added the second proviso describing the exemptions as "not exceeding five thousand gallons a day."

Furthermore, there is no indication that the enacting Legislature ever contemplated the possibility of the industrialization of the livestock industry. The Legislature may have anticipated that technology would eventually allow water use to exceed the Bureau of Reclamation's 1500 gpd estimation, but there is no evidence that the Legislature contemplated the possibility of future commercial livestock operations supporting thousands of cows and requiring as much as 100,000 gallons of water per day. Census statistics reveal that in the 1940s the average Washington farm was much smaller than today, and contemporary reports predicted livestock water requirements would be minimal for fully developed farms.

The Bureau of Reclamation's "vision of a rural West built on small farms" was in the process of being realized through the Columbia Basin Project, and the decline of the family farm and rise of industrialized, concentrated livestock farms did not begin occurring until several decades after the statute's enactment. The groundwater code was enacted to protect Washington's groundwater resources from the "destructive competition" and the "uncontrolled withdrawal of [ground]water" that had already caused damage to existing water rights and investments in the state, indicating that the Legislature's purpose in enacting the groundwater code was to prevent, not facilitate, unlimited withdrawals of groundwater. If the Legislature could have envisioned the possibility of concentrated livestock operations, it almost certainly would have specifically limited the stock-watering exemption.

Pre-enactment historical circumstances demonstrate that the 1945 Legislature most likely considered stock-watering to be naturally limited to small withdrawals, making an express limitation within the statute unnecessary. Post-enactment circumstances indicate that the Legislature agreed with the Department of Conservation's interpretation limiting all exempt withdrawals to 5000 gpd. There is no evidence suggesting the Legislature recognized or considered the possibility of the industrialization of the livestock industry that subsequently increased the amount of stock-water a single farm is capable of utilizing. These historical circumstances support the conclusion that the Legislature intended to limit all exempt uses, including stock-watering, to 5000 gpd.

The Public Policy Underlying the Groundwater Code Supports Limiting the Stock-watering Exemption to 5000 Gallons Per Day

Finally, contemporary groundwater management issues and the underlying policy of the groundwater code support a narrow interpretation of the stock-watering exemption. Although Washington courts have stated that it is the Legislature's duty to amend statutes that changing societal conditions have rendered obsolete, they have also said that where two reasonable interpretations of a statute exist, the interpretation that better advances the overall legislative purpose and underlying public policy should be adopted. In this case, no amendment is required because the text of the statute, related statutes, and historical context all indicate that the stock-watering exemption was intended to be limited when it was enacted.

Between the limited and unlimited interpretation, the limited interpretation better advances the groundwater code's underlying policy of "regulating and controlling ground waters of the state" and maintaining a "safe sustaining yield" of groundwater. Increasing population pressures, declining groundwater levels, and Ecology's severe backlog of permits and lack of adequate funding are currently impeding sustainable management of Washington's groundwater resources.

Declining groundwater levels are especially significant in the dry eastern region of the state, which also contains most of state's livestock industry. An unlimited stock-watering exemption could have a direct impact on the communities in eastern Washington that depend on groundwater. In an era of declining groundwater resources, the groundwater code's purpose of protecting senior water rights and managing groundwater in a sustainable manner is of primary importance. Allowing unlimited exemptions from the groundwater code's regulatory scheme undermines these important policy goals.

Washington's stock-watering exemption for groundwater is limited to 5000 gpd under Washington's principles of statutory interpretation. When interpreting statutes, the court's primary goal is ascertaining and implementing the Legislature's intent. The plain meaning and historical context of the exemption statute show that the Legislature intended to limit all exempt withdrawals to small withdrawals of 5000 gpd. Contemporary groundwater management issues and the groundwater code's underlying policy goal of managing Washington's groundwater resources in a sustainable manner reinforce this interpretation.

While this Comment is limited to a discussion of the proper interpretation of the current stock-watering exemption, disputes over the proper interpretation of the exemption statute will not be truly resolved until the discussion is shifted away from what legislators intended in 1945 to a discussion of the proper role for exempt withdrawals today. After years of encouraging both agricultural and urban growth, Washington must realize that water is a limited resource that cannot sustain all uses over time. Revising the groundwater code to limit exempt uses and enhance Ecology's ability to regulate the state's water resources will be necessary to support future urban and agricultural growth in the state.

Bibliography

Alka Rani Upadhyay: *Aquatic Plants for the Waste Water Treatment*, Daya, Delhi, 2004.

Anderson, Mary P. and William W. Woessner: *Applied Groundwater Modeling: Simulation of Flow and Advective Transport.* Academic Press, San Diego, 1992.

Asaithambi S. *: Economics of Ground Water Management in India*, Abhijeet, Delhi, 2008.

Asit K. Biswas and Cecilia Tortajada: *Appraising Sustainable Development: Water Management and Environmental Challenges*, Oxford University Press, Delhi, 2005.

Bagis, Ali Ihsan: *Water as an Element of Cooperation and Development in the Middle East*, Ankara, Ayna Publications, 1994.

Batu, Vedat: *Aquifer Hydraulics: A Comprehensive Guide to Hydrogeologic Data Analysis.* John Wiley & Sons, New York, 1998.

Bhakar S.R. *: Ground Water Hydrology : Theory and Practice*, Agrotech, Delhi, 2009.

Bhaskara, Harry: *Ending the Daily Water Chore In Against All Odds,* Panos Publications. London. 1989

Bhave, P.R. and R. Gupta: *Analysis of Water Distribution Networks*, Narosa, Delhi, 2011.

Biswas Asit K. *: Integrated Water Resources Management in South and South-East Asia*, , Oxford University Press, Delhi, 2001.

Bourne, Peter G.: *Water and Sanitation: Economic and Sociological Perspectives* Academic Press. Orlando. 1984.

Brooks, D.: *Water: Local-Level Management*, Ottawa, International Development Research Centre, 2002.

Bullock, John and Darwish, Adel: *Water Wars: Coming Conflicts in the Middle East*, London, Victor Gollancz, 1999.

Cazmuri, Renato : *"Chilean Water Policy Experience"*, Agriculture and Natural Resources Department, World Bank, Washington, D.C. Processed, 1992.

Conway, D.: *Climate Change and Water Resources in the Nile Basin*, London, University of London, 1993.

Cullis, A. : *Rainwater Harvesting: The Collection of Rainfall and Runoff in Rural Areas,* London, U.K.: IT Publications, 1986.

D'Itri, Frank M. and Lois G. Wolfson: *Rural Groundwater Contamination.* Lewis Publishers, Chelsea, 1987.

Domenico, Patrick A. and Franklin W. Schwartz: *Physical and Chemical Hydrogeology.* Wiley, New York, NY. 1990.

Erik Nissen-Peterson : *Rainwater Catchment Systems,* UK: Intermediate Technology Publications, 1999.

Falkenmark, Malin: *Rural Water Supply and Health: The need for a New Strategy,* Scandinavian Institute of African Studies. Uppsala. 1982.

Feder, Paul A.: *A Guidebook to Groundwater: Resources and Education Opportunities in the Great Lakes Region,* Great Lakes Commission, Ann Arbor, 1993.

Ferentinos L.: *Proceeding of the Sustainable Taro Culture for the Pacific Conference,* Honolulu, HITAHR, 1993.

Fetter, Charles Willard: *Applied Hydrogeology,* Merrill Publishing Company, Columbus, OH. 1988.

Freeze, R. Allan and John A. Cherry: *Groundwater.* Prentice Hall, Englewood Cliffs, NJ. 1979.

Geraghty, James J.: *Water Atlas of the United States,* Port Washington, NY, 1973.

Ghosh, N.C. and K.D. Sharma: *Groundwater Modelling and Management,* Capital Pub, Delhi, 2006.

Gischler, Christiaan E.: *Water Resources in the Arab Middle East and North Africa,* Cambridge, Eng, Middle East & North African Studies Press, 1979.

Goldman, Joanne Abel: *Building New York's Sewers: Developing Mechanisms of Urban Management,* West Lafayette, Ind.: Purdue University Press, 1997.

Gurjar, Ram Kumar : *Geography of Water Resources,* Rawat, Delhi, 2008.

Heath, Ralph C.: *Basic Ground-water Hydrology,* Denver, CO: U.S. Geological Survey, 1983.

Hemenway, Toby : *Gaia's Garden: A Guide to Home-Scale Permaculture,* Vermont: Chelsea Green Publishing Company, 2000.

Husain, Ahmad : *Environment and Water Resource Management,* Sumit Enterprises, Delhi, 2006.

Jana B L : *Water Harvesting and Watershed Management,* Agrotech, Delhi, 2008.

Jansky Libor : *Enhancing Participation and Governance in Water Resources Management: Conventional Approaches and Information Technology,* Bookwell, Delhi, 2006.

Jat, M.L. and S.R. Bhakar: *Ground Water Hydrology : Theory and Practice,* Agrotech, Delhi, 2009.

Jee Chandrawati and Shagufta : *Rainwater Harvesting,* A.P.H. Pub, Delhi, 2010.

Jeet Inder : *Rainwater Harvesting*, Mittal, Delhi, 2009.

Kally, Elisha and Gideon Fishelson: *Water and Peace: Water Resources and the Arab-Israeli Peace Process*, Westport, CT, Praeger, 1993.

Kanmony, J. Cyril : *Drinking Water Management : Problems and Prospects*, Mittal Pub, Delhi, 2010.

Khan M A : *Water Resources Management and Sustainable Agriculture*, APH, Delhi, 2008.

LeRoy, K. : *Sustaining Water Washington*, D.C.: Population Services International, New York, 1993.

Lloyd Myers : *Handbook of Water Harvesting*, Washington D.C.: U.S. Dept. of Agriculture, Agricultural Research Service, 1983

Mahajan Gautam : *Evaluation and Development of Ground Water*, APH, Delhi, 2008.

Matthess, Georg : *The Properties of Groundwater*, John Wiley & Sons, New York, NY. 1982.

McGarry, M.G.: *Matching Water Supply Technology to the Needs and Resources of Developing Countries,* United Nations. New York. 1987

Meenu Bhatnagar: *Groundwater Management : Sustainable Approaches*, Icfai Books, Delhi, 2012.

Merrett, Stephen, *Water for Agriculture: Irrigation Economics in International Perspective*, London, Spon Press, 2001,

Mishra Archana : *Water Harvesting : Ecological and Economic Appraisal*, Authors Press, Delhi, 2006.

Montgomery, John H. and Linda M. Welkom: *Groundwater Chemicals Desk Reference*, Chelsea, MI, 1990.

Morris, Mary E.: *Water Scarcity and Security concerns in the Middle East*, Abu Dhabi, Emirates Center for Strategic Studies and Research, 1998.

Murakami, Masahiro, *Managing Water for Peace in the Middle East : Alternative Strategies*, Tokyo and NYC, United Nations University Press, 1995.

Narasaiah, M. Lakshmi : *Energy, Irrigation and Water Supply*, Discovery, Delhi, 2004.

Narwani, G.S. : *Community Water Management*, Rawat, Delhi, 2005.

Nielsen, David M.: *Practical Handbook of Groundwater Monitoring*. Lewis Publishers, Chelsea, 1991.

Nissen-Petersen, E. : *Rainwater Catchment Systems for Domestic Supply: Design, Construction and Implementation,* London, U.K.: IT Publications, 1999.

Pahwa Prem S. : *Water Harvesting, Purification and Distribution Management*, Dominant, Delhi, 2001.

Palanisami, K : *Groundwater Management and Policies,* Macmillan Publishers India, Delhi, 2008.

Parikshit Ballabh: *Water Resource Management : Planning and Development*, Cyber Tech, Delhi, 2009.

Patrizio Warren: *Developing Participatory and Integrated Watershed Management*, Daya for FAO, Delhi, 2001.

Peter H. : *Water "Crisis: A Guide to the World's Fresh Water Resources"*, New York: Oxford University Press, 1933.

Postel, Sandra : *Last Oasis: Facing Water Scarcity*, World Water Institute, New York, 1992.

Premjit Sharma: *Agricultural Drainage and Water Quality*, Gene Tech Books, Delhi, 2007.

Ranghaswami, M.V. ; K. Palanisami: *Groundwater Resources Assessment, Recharge and Modelling*, Macmillan Publishers India, Delhi, 2008.

Rao K. Nageswara : *Water Resources Management : Realities and Challenges*, New Century Publication, Delhi, 2006.

Reddy, K L N : *Economics of Rural Drinking Water Supply*, University Book House, Delhi, 1999.

Rowe, Garry W. and Sylvia J. Dulaney: *Building and using a Groundwater Database*. Lewis Publishers, Chelsea, 1991.

Salameh, Elias and Helen Bannayan: *Water Resources of Jordan: Present Status and Future Potentials*, Amman, Friedrich-Ebert-Stiftung, 1993.

Sharma, R. K. and T.K.Sharma: *A Textbook Of Water Power Engineering*, S. Chand Publisher, Delhi, 2008.

Sharma, S.S.P. and U.H. Kumar: *Dynamics of Watershed Development and Livelihood in India*, Serials Pub, Delhi, 2010.

Singh A K : *Environment and Water Resources Management*, Adhyayan, Delhi, 2006.

Thapliyal, B K : *Democratisation of Water*, Serials Pub, Delhi, 2008.

Thomas, Harold E.: *The Conservation of Groundwater: A Survey of the Present Groundwater Situation in the U.S.*, McGraw Hill, New York, NY. 1951.

Todd, David Keith: *Groundwater Hydrology*. John Wiley & Sons, New York, NY. 1959.

Van der Leeden, Frits: *Ground Water: A Selected Bibliography*, Port Washington, NY, 1974.

Walton, William C.: *Groundwater Resource Evaluation*. McGraw Hill, New York, 1970.

William, K. : *Irrigation Investment Technology, and Management Strategies for Development Studies in Water Policy Management 9*, Boulder, Colo Westview Press, 1986.

Yaca Tsur : *"Water Shadow Values and Institutional Arrangements for Allocating Water among Competing Sectors*', Department of Agricultural and Applied Economics, University of Minnesota. Processed, 1992.

Index

R

S

T

V

W

□□□